Socialising Tourism

Once touted as the world's largest industry and also a tool for fostering peace and global understanding, tourism has certainly been a major force shaping our world. The recent COVID-19 crisis has led to calls to transform tourism and reset it along more ethical and sustainable lines. It was in this context that calls to "socialise tourism" emerged (Higgins-Desbiolles, 2020). This edited volume builds on this work by employing the term *Socialising Tourism* as a broad conceptual focal point and guiding term for industry, activists and academics to rethink tourism for social and ecological justice.

Socialising Tourism means reorienting travel and tourism based on the rights, interests and safeguarding of traditional ecological and cultural knowledges of local peoples, communities and living landscapes. This means making tourism work for the public good and taking seriously the idea of putting the social and ecological before profit and growth as the world re-emerges from the COVID-19 pandemic. This is an essential first step for tourism to be made accountable to the limits of the planet. Concepts discussed include Indigenous culture, toxic tourism, a "theory of care", dismantling whiteness, decolonial tourism and animal oppression, among others, all in the context of a post-COVID-19 world.

This will be essential reading for all upper-level students, academics and policymakers in the field of tourism.

Freya Higgins-Desbiolles is a Senior Lecturer in Tourism Management, UniSA Business, University of South Australia and adjunct Associate Professor, Department of Recreation and Leisure, University of Waterloo, Canada. She has worked with communities, non-governmental organisations and businesses that seek to harness tourism for sustainable and equitable futures. She is one of the Founding members of the Tourism Alert and Action Forum. She has won awards for engaged research, media engagement and research and teaching excellence.

Adam Doering is an Associate Professor in the Faculty of Tourism at Wakayama University, Japan. He is a Steering Committee Member of the Critical Tourism Studies Asia Pacific (CTS-AP) research network and has published broadly on the philosophy and ethics in tourism and travel, lifestyle sports and tourism in East Asia and critical analyses of Destination Management Organisations policies in the context of Japan. His current research examines lifestyle sports and tourism development in polluted, post-disaster and pandemic impacted coastal ecologies in rural Japan.

Bobbie Chew Bigby (Cherokee Nation) is a PhD student at the Nulungu Research Institute, University of Notre Dame Australia. Bobbie's research looks at the possibilities of tourism as a tool for Indigenous cultural, language and environmental justice. Her past research fellowships, including a Fulbright award and Rotary Peace Fellowship, have taken her to Indigenous Australia, China, India, Cambodia and Burma for research and community-based work. For her PhD research, Bobbie rotates her time between Broome, Western Australia and her own Tribal Nation in Oklahoma, USA.

Contemporary Geographies of Leisure, Tourism and Mobility

Series Editor: C. Michael Hall, *Professor at the Department of Management, College of Business and Economics, University of Canterbury, Christchurch, New Zealand*

The aim of this series is to explore and communicate the intersections and relationships between leisure, tourism and human mobility within the social sciences.

It will incorporate both traditional and new perspectives on leisure and tourism from contemporary geography, e.g. notions of identity, representation and culture, while also providing for perspectives from cognate areas, such as anthropology, cultural studies, gastronomy and food studies, marketing, policy studies and political economy, regional and urban planning and sociology, within the development of an integrated field of leisure and tourism studies.

Also, increasingly, tourism and leisure are regarded as steps in a continuum of human mobility. Inclusion of mobility in the series offers the prospect to examine the relationship between tourism and migration, the sojourner, educational travel and second home and retirement travel phenomena.

The series comprises two strands:

Contemporary Geographies of Leisure, Tourism and Mobility aims to address the needs of students and academics, and the titles will be published simultaneously in hardback and paperback.

Routledge Studies in Contemporary Geographies of Leisure, Tourism and Mobility is a forum for innovative new research intended for research students and academics, and the titles will initially be available in hardback only. Titles include:

Socialising Tourism
Rethinking Tourism for Social and Ecological Justice
Edited by Freya Higgins-Desbiolles, Adam Doering and Bobbie Chew Bigby

For more information about this series, please visit: www.routledge.com/Contemporary-Geographies-of-Leisure-Tourism-and-Mobility/book-series/SE0522

Socialising Tourism

Rethinking Tourism for Social and Ecological Justice

Edited by
Freya Higgins-Desbiolles,
Adam Doering and
Bobbie Chew Bigby

LONDON AND NEW YORK

First published 2022
by Routledge
2 Park Square, Milton Park, Abingdon, Oxon OX14 4RN

and by Routledge
605 Third Avenue, New York, NY 10158

Routledge is an imprint of the Taylor & Francis Group, an informa business

British Library Cataloguing-in-Publication Data
A catalogue record for this book is available from the British Library

Library of Congress Cataloging-in-Publication Data
Names: Higgins-Desbiolles, Freya, editor. | Doering, Adam, editor. | Chew Bigby, Bobbie, editor.
Title: Socialising tourism : rethinking tourism for social and ecological justice / edited by Freya Higgins-Desbiolles, Adam Doering and Bobbie Chew Bigby.
Description: Abingdon, Oxon ; New York, NY : Routledge, 2022. | Series: Contemporary geographies of leisure, tourism and mobility | Includes bibliographical references and index.
Identifiers: LCCN 2021012825 (print) | LCCN 2021012826 (ebook)
Subjects: LCSH: Tourism—Social aspects. | Tourism—Environmental aspects. | Social justice. | Environmental justice.
Classification: LCC G156.5.S63 S65 2022 (print) | LCC G156.5.S63 (ebook) | DDC 306.4/819—dc23
LC record available at https://lccn.loc.gov/2021012825
LC ebook record available at https://lccn.loc.gov/2021012826

ISBN: 978-0-367-75922-3 (hbk)
ISBN: 978-0-367-75925-4 (pbk)
ISBN: 978-1-003-16461-6 (ebk)

DOI: 10.4324/9781003164616

Typeset in Times New Roman
by codeMantra

Bobbie Chew Bigby

Gratitude to my loved ones and elders who have shown me unwavering support in all aspects of my journey. I am particularly thankful to my co-editing team, Freya and Adam, who have been exceptional mentors and have encouraged me to keep dreaming big and moving forward.

I dedicate this book to the memory of all the lives lost due to the COVID-19 pandemic as it has touched all corners of this planet we call home.

Freya Higgins-Desbiolles

I am grateful for all of those strong voices who have championed justice and human rights in tourism through many decades, some of whom are included in this collection. I also hope this book will serve as inspiration to those who will continue the struggle.

Adam Doering

I dedicate this book to the people of Fukushima whose kindness and compassion in the face of ongoing adversity are truly inspirational.

Contents

Figures

Contributors

Editors

Bobbie Chew Bigby (Cherokee Nation) is a PhD student at the Nulungu Research Institute, University of Notre Dame Australia. Bobbie's research looks at the possibilities of tourism as a tool for Indigenous cultural, language and environmental justice. Her past research fellowships, including a Fulbright award and Rotary Peace Fellowship, have taken her to Indigenous Australia, China, India, Cambodia and Burma for research and community-based work. For her PhD research, Bobbie rotates her time between Broome, Western Australia and her own Tribal Nation in Oklahoma, USA.

Adam Doering is an Associate Professor in the Faculty of Tourism at Wakayama University, Japan. He is a Steering Committee Member of the Critical Tourism Studies Asia Pacific (CTS-AP) research network and has published broadly on the philosophy and ethics in tourism and travel, lifestyle sports and tourism in East Asia and critical analyses of Destination Management Organisations policies in the context of Japan. His current research examines lifestyle sports and tourism development in polluted, post-disaster and pandemic impacted coastal ecologies in rural Japan.

Freya Higgins-Desbiolles is a Senior Lecturer in Tourism Management, UniSA Business, University of South Australia and adjunct Associate Professor, Department of Recreation and Leisure, University of Waterloo, Canada. She has worked with communities, non-governmental organisations and businesses that seek to harness tourism for sustainable and equitable futures. She is one of the Founding members of the Tourism Alert and Action Forum. She has won awards for engaged research, media engagement and research and teaching excellence.

Authors

Stefanie Benjamin, PhD, is an Assistant Professor in the Retail, Hospitality and Tourism Management Department at the University of Tennessee. Her research interests include social equity in tourism around the intersectionality of race, gender, sexual orientation and people with disabilities. She is also the Co-Director of Tourism RESET. She researches film-induced tourism, implements improvisational theatre games as innovative pedagogy and is a certified qualitative researcher exploring ethnography, visual methodology and social media analysis.

Raoul V. Bianchi is Reader in International Business in the Department of Economics, Policy and International Business & Future Economies University Research at Manchester Metropolitan University. His work focuses on the political economy and politics of international tourism, tourism geopolitics and citizenship, tourism work and labour relations, sustainability and the transition to tourism economies beyond capitalism. His principal geographical area of interest lies in southern Europe and the Mediterranean. He is currently an Associate Editor of *Annals of Tourism Research: Empirical Insights* and previously sat on the executive council of the former UK NGO Tourism Concern.

Asunción Blanco-Romero is Associate Professor in the Department of Geography at the Universitat Autònoma de Barcelona (Spain), and also participates with UNED (National University of Distance Education), UOC (Universitat Oberta de Catalunya) and OSTELEA School of Tourism and Hospitality. She is a member of TUDISTAR-UAB research group (Tourism and new social and territorial dynamics), with which she has participated in several research projects. Her research focuses on tourisms and local development, tourism and degrowth and geography and gender issues in regional development.

Macià Blázquez-Salom is Associate Professor in the Department of Geography at the University of the Balearic Islands. His research interests include tourism territorial planning and nature conservation from a sustainability perspective. He has been a visiting scholar in several European and Central American universities. His most recent publications deal with urban and regional planning regulation and the expansion of the Balearic Islands' hotel chains in Central America and the Caribbean.

Karla Boluk, PhD, is an Associate Professor in the Department of Recreation and Leisure Studies at the University of Waterloo in Canada. Utilising a critical lens, Karla's scholarship investigates ways to sustainably engage and empower communities, positioning tourism as a mechanism for the creation of positive change. Her main research interests explore the role of tourism social entrepreneurs in progressing sustainability, specifically the SDGs, and investigating ways to activate critical thinking in students, preparing graduates as critical tourism change agents.

Ernest Cañada is the founder and coordinator of Alba Sud, an independent research centre specialised in tourism from critical perspectives. He also collaborates as an associate professor at the University of Barcelona in Spain and at the University of Angers in France. Recently he has published, together with Ivan Murray, a book on global touristification. Among his lines of research, the following stand out: (a) studies on work in tourism, (b) socio-ecological conflicts linked to tourism, (c) community tourism and (d) post-capitalist alternatives in the organisation of tourism.

Sandro Carnicelli is Senior Lecturer at the University of the West of Scotland and the Unit Leader for the Marketing, Events and Tourism group. Sandro has been developing research and working in the fields of Tourism and Events for over 15 years and his main research interests are: tourism education; critical pedagogy; adventure tourism; emotional labour; and outdoor learning. Sandro is currently a member of the editorial board of *Leisure Studies* and associate editor of the *Journal of Adventure Education and Outdoor Learning.*

Natasha Chassagne is an Adjunct Research Fellow at Centre for Social Impact, Swinburne University of Technology, Australia. She holds a PhD looking at Buen Vivir as an Alternative to Sustainable Development. Dr. Chassagne has recently published a book on the same topic.

Alana Dillette, PhD, is an Assistant Professor in the Payne School of Hospitality and Tourism Management as well as the Co-Director for Tourism RESET, an initiative focused on race, ethnicity and social equity in tourism. Dr. Dillette also serves as a liaison for diversity initiatives in the College of Professional Studies and Fine Arts. Originally from the islands of The Bahamas, Dr. Dillette conducts research that explores the intersection between tourism, race, gender and ethnicity. More specifically, she is working on research to gain a better understanding of the Black travel experience in addition to the challenges faced by Black hospitality and tourism professionals.

Phoebe Everingham is an early career researcher and sessional staff member at the University of Newcastle, Australia. She draws on multidisciplinary perspectives such as human geography, sociology, anthropology and tourism management studies.

Robert Fletcher is Associate Professor in the Sociology of Development and Change group at Wageningen University in the Netherlands. His research interests include tourism, conservation, development, climate change and resistance and social movements.

Rebecca Jim holds a BA in Behavioral Sciences from Southern Colorado State College and an MA in Education and Counseling from Northeastern State University. She is a Co-Founder and Executive Director of LEAD (Local Environmental Action Demanded) Agency, a non-profit organisation focused on environmental justice for northeast Oklahoma. She is also a member of the Cherokee Nation.

Kyle Kajihiro is an American Council of Learned Societies Emerging Voices Postdoctoral Fellow at the University of Wisconsin Madison. He has an MA and a PhD in Geography from the University of Hawai'i at Mānoa, where he also lectures in Ethnic Studies and Geography and Environment. From 1996 to 2011 he was the Programme Director for the American Friends Service Committee – Hawai'i Area Programme, where his work included research, public education and organising to advance environmental justice, solidarity with Kanaka 'Ōiwi and demilitarisation. This work led to the formation of the Hawai'i DeTours Project, a political education programme that examines the decolonial historical geography of Hawai'i.

Kumi Kato is a Professor in the Faculty of Tourism, Wakayama University, and a special appointed professor at Musashino University, Japan. She is also a board member of the Global Sustainable Tourism Council. Kumi has been leading a national project of development and implementation of Japan Sustainable Tourism Standard for Destinations for the Japan Tourism Agency, Ministry of Land, Infrastructure, Transport and Tourism, Japan. Her research area includes sustainability, community resilience, ethics and the eco-humanities in relation to tourism.

Carol Kline is an Associate Professor of Hospitality and Tourism Management at Appalachian State University. Her teaching and research interests have historically focused broadly on tourism sustainability, including topics such as foodie segmentation, craft beverages, agritourism, tourism entrepreneurship and tourism in developing economies. However, she now gears her research solely on animals and she teaches a course called Animals, Tourism and Sustainability. She is part of the Race, Ethnicity and Social Equity in Tourism (RESET) initiative, which includes animals within the study of social equity. She is founder of Fanimal Inc., a non-profit that helps individuals find animal-focused careers.

Simon Lambert is from Aotearoa New Zealand and is a member of the Tuhoe and Ngati Ruapani tribes. His doctoral research was on small-scale Maori horticulture and he was awarded his PhD in 2008 from Lincoln University. Prior to that, he was awarded a MA (Hons) degree from the University of Canterbury after researching the assessment of environmental vulnerability in the Pacific. Following the 2010–11 earthquakes in Christchurch, Simon's research has focused on disaster risk reduction for Indigenous communities with particular interest in urban Indigenous groups.

Deborah McLaren holds a MA in Social Ecology from Goddard College, specialising in community-based tourism in Asia. She was an Asia Foundation Fellow in ecotourism research and collaboration in Sikkim, India. She served as director of two international non-profit organisations focused on sustainable tourism and justice, the Rethinking Tourism Project (1996–2001) and Indigenous Tourism Rights International (2001–2005). During this work, she helped coordinate the first International Indigenous Tourism Forum in Oaxaca, Mexico in 2003 and the Indigenous Tourism Certification Conference in 2004. Deborah is the author of *Rethinking Tourism and Ecotravel* (1998; 2003), publications influential in the development of the ethics of tourism. Deborah served as a UNDP consultant to the government of Bhutan and in the development of "Journeys with First Nations", a Green Route on White Earth Nation in Minnesota, USA. Currently she serves on the steering committee of the Indigenous Tourism of the Americas Collaborative. She owns an artisanal spice company in Minnesota.

Kokel Melubo is a Senior Lecturer at the College of African Wildlife Management, Mweka, Tanzania. He received his PhD in Tourism from the University of Otago, New Zealand, and completed an MS in Natural Resources Management and Bachelor of Arts (Hons) degree in Geography from the University of Dar es Salaam. His current research and publications include the areas of Indigenous tourism, sustainable tourism, visitor management and tour guiding in the Tanzanian context.

Rasul A. Mowatt is a Professor in the Departments of American Studies and Geography in the College of Arts and Science, Indiana University. His primary areas of research are: Geographies of Race, Geographies of Violence and Threat, The Animation of Public Space and Critical Leisure Studies. His published work has been on analysing racially violent forms of leisure in the *American Behavioral Scientist*, the dangers in viewing images of Black death in *Biography* and most recently, the threat of violence from intimate terrorism, White Nationalists and the State in a special call for COVID-19 in *Leisure Sciences*.

Ivan Murray Mas holds a PhD in Geography from the University of the Balearic Islands and an MsC in Environmental Sustainability from the University of Edinburgh. He is Associate Professor of geography at the University of the Balearic Islands. He is a member of the Research Group on Sustainability and Space. His research merges political ecology, political economy and ecological economics of tourism. He has recently co-edited a book on global touristification with Ernest Cañada. He is also involved in social movements.

Andrew Peters is Senior Lecturer in Indigenous Studies and Tourism at Swinburne University of Technology and a descendent of Wundjeri/Yarra Yarra, Yorta Yorta and Ngurai Illum Wurrung peoples from Victoria, Australia. Andrew's research interests include contemporary cultural identity, sport and culture, Indigenous tourism development and contemporary utilisation of Indigenous Knowledges. Recent projects include an evaluation of a study support programme for Indigenous inmates at Port Phillip Prison and an evaluation of cultural programmes for the Korin Gamadji Institute at the Richmond Football Club.

Filka Sekulova is a postdoctoral researcher at the Institute of Environmental Science and Technology (ICTA) at the Universitat Autònoma de Barcelona, with a background in ecological economics, psychology and urban environmental justice.

Shinji Yamashita is Emeritus Professor of Cultural Anthropology at the University of Tokyo, Japan. He served as President of the Japanese Society of Ethnology (currently the Japanese Society of Cultural Anthropology) from 1996 to 1998 and President of the Japan Society of the Interdisciplinary Tourism Studies from 2012 to 2019. He is also a Fellow of the International Academy for the Study of Tourism. His research and publication output focuses mainly on the analysis of cultural dynamism in relation to tourism and migration in Asia-Pacific region.

Foreword

When will change come?

Deborah McLaren

> In the midst of every crisis, lies great opportunity.
>
> Albert Einstein

Different perspectives

This famous quotation by Albert Einstein seems timely when thinking about the prospects of tourism businesses surviving through the pandemic and confronting the challenges global climate change is bringing. Of course, it also depends on the perspective from which you are considering it. Throughout the history of travel, one is struck by the differing views of land and place held by travellers and hosts, colonisers and original inhabitants, and the compulsion of the newcomers to take control of their "discoveries", including the resources and people found there. Many rural and Indigenous communities have been displaced not only from their lands but also from their traditional cultures, trades and livelihoods. Often they have been compelled to work for less sustenance in the service industry than they would have enjoyed if their subsistence lifeways had been allowed to continue on as before. This occurs as they watch their lands become centres of high energy consumption, senseless waste, meaningless consumerism and now, with the COVID-19 pandemic, disease. The essential facts are that the tourism industry has become a central component of economic globalisation, climate change and now viral pandemics.

I have been an advocate of sustainable tourism for more than three decades and I was asked to write the foreword for this edited book. My insights are derived from serving as the director of two international non-profits focused on sustainable tourism and justice, the Rethinking Tourism Project and Indigenous Tourism Rights International. During this work, I helped coordinate the first International Indigenous Tourism Forum in Oaxaca, Mexico in 2003 and the Indigenous Tourism Certification Conference in 2004. I am the author of *Rethinking Tourism and Ecotravel* (1998; 2003) as well as numerous magazine and online articles. Additionally, I have served as a consultant to the government of Bhutan and in the development of "Journeys with First Nations", a Green Route on White Earth Nation in Minnesota. This body of work has been a collaborative effort alongside non-profits, communities, small businesses, tribes and government agencies, with a commitment to designing culturally appropriate, sustainable tourism initiatives. My career is marked by continual work on critical theoretical analysis integrated with reflective praxis that is engaged, collaborative and responsive to those forced to live with and through tourism.

It is from this background that I present the provocation: when will change come?

Coronavirus and global pandemics

There have been numerous articles, reports and presentations about the impacts of the COVID-19 crisis. The crisis has upended our societies through the measures needed to address it, including shutting borders and stopping tourism in its many forms. While some aid has been allocated to local communities for help in light of COVID-19, it is primarily to support training to meet health and safety guidelines. This reflects their focus on diminishing the crisis and return to "normal" as soon as possible in order to re-establish mainstream development processes when the pandemic passes. Yet, if we've learned anything from this, shouldn't we be planning for future epidemics and other crises which are clearly on their way? If so, more than just health and safety training are required.

Many communities have discovered how completely reliant they have become on the tourism and hospitality industries. Immense suffering has occurred without the jobs and income people have come to rely on in tourism-dependent communities around the world. The COVID-19 crisis has forced communities to accept the need to change or otherwise perish. Some are embracing innovation and adopting new tourism methods and techniques with a view to delivering sustainable, efficient, affordable and meaningful experiences for domestic visitors in an effort to survive and indeed flourish. However, we cannot overlook that some communities have decided to return to more traditional undertakings to enhance food security and environmental sustainability (see Laula & Paddock, 2020, on Bali; Scheyvens & Movono, 2020, on the Pacific).

Indigenous experiences illuminate important considerations in relation to these critical reflections on tourism. The pandemic poses a grave health threat to Indigenous peoples around the world. Indigenous communities already experience poor access to healthcare, significantly higher rates of communicable and non-communicable diseases, lack of access to essential services, food shortages, loss of livelihoods, unemployment, little or no government relief, debt and exacerbation of poverty. Isolation has also brought violence and intimidation to many communities as individual land grabbers, private companies and security forces are continuing to occupy the territories of Indigenous communities and/or take advantage of the crisis.

Indigenous ceremonies, farming and community gatherings are often large with everyone participating, which further puts them at risk. In fact, tourism has long been known to bring illness to many Indigenous nations who are not protected from outside diseases with immunity. They are still determining ways to protect their intellectual property rights and cultural property. They are also using traditional knowledge and practices to take action, such as voluntary isolation and sealing off their territories, along with other preventative measures. Tourism – and any current incoming aid – especially amid the pandemic, is still not addressing the existing systemic and structural human rights issues that continue to marginalise Indigenous communities.

Victoria Tauli-Corpuz, UN Special Rapporteur on the Rights of Indigenous Peoples, calls on us to use the pandemic for the good of humanity:

> COVID-19 has shown us that atomized societies which put individualism and profit-making as the pinnacle of modern civilization will further lead us to more disasters and crises. If people love and have more compassion, empathy and solidarity for and with other human beings and nature, our world will not be in this state (2020).

This is the opportunity that the pandemic crisis affords us. It potentially socialises us to eschew greediness, exploitation and imposition and choose pathways to greater solidarity, including in and through tourism. Such a transformation is essential as even greater crises are upon us.

Climate change

Arguably, the most important issue of our time, not discounting the spread of the coronavirus, is global climate change and its myriad of impacts. The Brookings Institute, however, noted that the increased frequency and severity of disease outbreaks are caused by climate change (Podesta, 2019); clearly the two issues are intertwined.

The tourism industry is implicated in these concerns; it has contributed to carbon in our atmosphere, placed greater demands on limited natural resources and created mountains of waste, in addition to numerous other problems. Our atmosphere is changing and that is reflected in the growing environmental disasters we face. Scientists predict that there will be mass migrations as coastal areas erode and other environmental crises, like major wildfires, will impact us. This isn't going to hit us later; it is hitting us now. The Arctic has been opened for cruise passages through areas that were previously frozen solid. The loss of geographical land is limiting our living space and the question is: who can afford to live in the space that will be habitable? Can we truly predict the tourism industry when faced with such a disastrous future? Will it boil down to restricting our freedom and consumption? Or will only those who can afford it continue to live in and vacation in the best places available? And whose lands are those? Although not always obvious, tourism is bound up with climate change, massive migration by environmental refugees and the resultant land-grabbing that will follow.

There are a few examples where migration strategies are included in recent agreements such as the 2015 Paris Agreement that have called on countries to make plans to prevent the need for climate-caused relocation and support those forced to relocate (United Nations, 2018).[1] However, these agreements are neither legally binding nor sufficiently developed to support climate migrants.

As gradually worsening climate patterns and severe weather events prompt an increase in human mobility, people who choose to move will do so with little legal protection. The current system of international law is not equipped to protect climate migrants, as there are no legally binding agreements obliging countries to support climate migrants (Ionesco, 2019). At the same time, where they are forced to move is of vital importance, as biologically diverse areas are the homelands of people who often have few rights.

Therefore, land recognition and legal ownership are serious matters for communities at the forefront of these issues. Insecure land tenure affects the ability of people, communities and organisations to make changes to land that can advance adaptation and mitigation. Land policies can provide both security and flexibility for responding to climate change, ensuring rural and Indigenous peoples can implement their own strategies for addressing climate change. While tourism is only one out of a myriad of players that have a role impacting such issues, it must be “socialised” for supporting life-enhancing outcomes rather than undermining them.

Resistance and change

For over three decades I’ve observed and written about tourism and globalisation and their community impacts. While there has been some change – such as recognition by some states and global development organisations – and the communities who live in these places have kept them fairly pristine and they are pretty good managers of their own ecosystems, there is still not enough local control and enforced guidelines. The idea of culture plays an increasingly important role in the concept of development, in part because the inherent inequalities have been challenged from the people themselves. Increasingly, this involves powerful assertions from Indigenous peoples and a new focus around issues of environmental sustainability of food sovereignty, land rights and damage caused by global environmental crisis, among others. All of these issues are bound up with climate change and collectively they represent a tremendous challenge for Indigenous peoples. As the International Working Group for Indigenous Affairs explained:

> Intensifying global competition over natural resources increasingly makes Indigenous communities, who act to protect their traditional lands and territories, targets of persecution by State and non-State actors. Recent reports by UN experts and human rights organisations document an alarming increase in violent attacks against and

> criminalisation of indigenous peoples defending their rights to their traditional lands and natural resources.
>
> (IWGIA, 2019)

In October 2020, the International Forum on Indigenous Tourism took place online. Before the spread of COVID-19, tourism provided a promising approach to sustainable development – as Indigenous leaders pointed out – a way for Indigenous communities to generate income, alleviate poverty, increase access to healthcare and education, and conserve their cultural and natural resources. After more than 25 years of organising their communities locally, nationally and internationally, Indigenous peoples have developed and are re-establishing ancient networks, exchanging knowledge, sharing reflections and finding their way forward so that Indigenous people everywhere can continue to be empowered to determine their own fates and futures.

I participated in several sessions of the International Forum on Indigenous Tourism. I noted that while the global coronavirus pandemic has virtually put a stop to most types of tourism, it has created a desire for many people who have been isolating under forms of pandemic lockdowns to get out into nature and be in the wild – often attracted to Indigenous lands or their homelands (such as protected areas). Indigenous tourism leaders at the forum discussed emerging issues, namely that the tourism economy that they are part of has greatly decreased as there are more individuals and smaller groups travelling on their own without local guidance and support services. This results in less income, less control and more health and safety risks. Some local travel companies have taken the opportunity to reassess their tourism businesses and determine how to best align them with their moral and ethical values. For too long they have found themselves managing a growing industry and often feeling as though they have disregarded their own cosmovision and traditions.

Several laws and concepts related to tourism are potentially shaping a more just concept of tourism for all, including human rights protocols, land rights, land acknowledgement, reconciliation and social justice provisions. For visitors and tourism businesses, it is important to acknowledge whose lands you are visiting or operating on. More importantly, it is vital to understand and respect local contexts and rights (including international, national-states and local rules and protocols). In terms of Indigenous peoples and tourism, the United Nations Declaration on the Rights of Indigenous Peoples (UNDRIP) warns that the failure to ensure land rights constitutes the core underlying cause of violations of Indigenous peoples' rights and affirms the right of Indigenous peoples to own and control their lands and territories in Articles 25, 26, 27 and 32 (UN, 2007). In the Declaration on Human Rights Defenders, it is stated that States should provide specific protections to human rights defenders and have the responsibility to implement and respect all the provisions of the Declaration (UN Human Rights, 1998).

One example of leadership in reorienting tourism and socialising it to local contexts and needs can be found in the United States. Through the persuasive efforts of US tribes, tribal organisations and the tourism industry, the United States Senate and House of Representatives voted to enact the Native American Tourism and Improving Visitor Experience (NATIVE) Act (*Public Law 114–221*); it was signed into law in 2016. It serves to establish a more inclusive national travel and tourism strategy and has the potential to deliver significant benefits for tribes, including job creation, elevated living standards, expanded economic opportunities and pathways to more self-determining futures. Importantly, it is administered by the American Indian and Alaska Native Tourism Association (AIANTA, n.d.). Numerous countries are enacting similar policies and strategies.

Locally, it is extremely important to understand and acknowledge *land rights*. That means really getting to know the current realities for rural and Indigenous communities. Whether a visitor, tourism business or government official, this is a priority. This includes the Indigenous peoples to whom the land belongs, the history of the land and any related treaties, Indigenous place names and language and correct pronunciation for the names of the tribes, places and individuals that are important.

However, the Native American-led non-profit the Native Governance Centre (NGC) asserts that *land acknowledgement* alone is not enough (NGC, 2019). It's merely a starting point. Both tourists and non-Indigenous tourism businesses can determine a plan to take action to support Indigenous communities under their guidance. Possible actions include:

- Support Indigenous organisations by donating time and/or money.
- Support Indigenous-led grassroots change movements and campaigns.
- Commit to returning land or supporting the return of land. Local, state and federal governments around the world are currently returning land to Indigenous Peoples. Individuals are returning their land, too. Learn more about options to return land.
- Identify land acknowledgement in the designated tourism area. If there isn't one, consider reaching out to local Indigenous leadership and local government to engage Indigenous peoples in the area to create one (NGC, 2019).

Reconciliation is the process of developing a respectful relationship between Indigenous peoples and settlers. It's about working together to overcome the devastating effects of colonisation. With reconciliation, it's important to acknowledge harmful policies and practices (e.g. residential schools, loss of lands, inequitable access to essential services, prohibition of cultural traditions and languages, land displacement etc.) and define positive ways to move forward together.

Tourism companies are currently working with rural and Indigenous Peoples on reconciliation projects. The strategy is to understand current issues and address adversity and inclusion in training, supply chains and business, and to interweave that work with *human rights* and *social justice* issues, especially the climate crisis. Creating a diverse, energetic working group to drive the agenda must include leaders and First Nations' representatives. This discussion has focused particularly on Indigenous interfaces with tourism because that is the domain of my work; however, these principles and practices are equally relevant to engaging with the marginalised, minorities, rural and peripheral communities, disaster effected areas, polluted landscapes, animals and the non-human world and other groups addressed in this edited volume that are impacted by tourism but also have the capacity to socialise tourism for better outcomes.

Socialising tourism

I have observed and supported many different communities and organisations that have sought to rethink and challenge mass tourism over three decades. When the invitation arose to review the concept of socialising tourism and open this book, I was enthusiastic, considering the context in which it has arisen amid the pandemic raising questions of the social, the marginalised and ways forward. Higgins-Desbiolles has described socialising tourism "as placing tourism in the context of the society in which it occurs and to harness it for the empowerment and wellbeing of local communities" (2020, p. 618). In this work, she advocated for a public good form of tourism, arguing that public expenditure supporting tourism must be used to support people's well-being and ecological restoration (Higgins-Desbiolles, 2020). This seems inherent to the logic of community-based tourism, the focus of my work for decades. In light of the crises I have outlined here, the need for change is clear and pressing. Placing the needs, rights and benefits of women, marginalised peoples, rural communities, Indigenous peoples and living ecosystems at the forefront of our tourism decisions and activities is the next step in the long struggle many of us have waged to tame tourism and harness it for building better futures.

I commend the chapters in this book for your reading. In the following book, well-respected authors have taken up the challenge to explore, critique and further our understanding of what socialising tourism might offer. Their work is structured to address three critical concerns: socialising tourism as rethinking social relations; socialising tourism as rethinking ideology; and socialising tourism to build better collective futures. There is no other book on the academic market like this one and it responds well to the challenging times we are confronting. It might not give the precise answer to the question "when will change come", but it certainly moves us closer down that road.

Note

1 Specifically, articles 18.H (share information to better map and predict migration based on climate change and environmental degradation), 18.I (develop adaptation and resilience strategies that prioritize the country of origin), 18.J

(factor in human displacement in disaster preparedness strategies), and 18.K (support climate-displaced persons at the subregional and regional levels) (United Nations, 2018).

References

AIANTA (n.d.). Understanding the Native Act, brochure. Retrieved 3 January 2021, from https://www.aianta.org/wp-content/uploads/2018/09/About-the-NATIVE-Act-Final.pdf

Higgins-Desbiolles, F. (2020). Socialising tourism for social and ecological justice after COVID-19. *Tourism Geographies, 22*(3), 610–623. https://doi.org/10.1080/14616688.2020.1757748

Ionesco, D. (2019). Let's talk about climate migrants, not climate refugees. Retrieved 6 January 2021, from https://www.un.org/sustainabledevelopment/blog/2019/06/lets-talk-about-climate-migrants-not-climate-refugees/

IWGIA (2019). Indigenous rights defenders at risk. Retrieved 27 November 2020, from https://www.iwgia.org/en/focus/indigenous-rights-defenders-at-risk.html.

Laula, N., & Paddock, R.C. (2020, July 20). With tourists gone, Bali workers return to farms and fishing. *New York Times* (Online). Retrieved 6 January 2020, from https://www.nytimes.com/2020/07/20/world/asia/bali-tourism-coronavirus.html

McLaren, D. (1998). *Rethinking tourism and ecotravel* (1st ed.). Bloomfield, CT: Kumarian.

McLaren, D. (2003). *Rethinking tourism and ecotravel* (2nd ed.). Bloomfield, CT: Kumarian.

NativeGovernanceCentre(2019).Indigenouslandacknowledgement.Retrieved6January2021,fromhttps://nativegov.org/a-guide-to-indigenous-land-acknowledgment/

Podesta, J. (2019). The climate crisis, migration, and refugees. *The Brookings Institute.* Retrieved 27 November 2020, fromhttps://www.brookings.edu/research/the-climate-crisis-migration-and-refugees/

Scheyvens, R., & Movono, A. (2020). Traditional skills help people on the tourism-deprived Pacific Islands survive the pandemic. *The Conversation.* Retrieved 11 January 2020, from https://theconversation.com/traditional-skills-help-people-on-the-tourism-deprived-pacific-islands-survive-the-pandemic-148987

Tauli-Corpuz, V. (2020, 16 April). Statement on COVID 19 and indigenous peoples UN special rapporteur on the rights of indigenous peoples. *International Working Group on Indigenous Affairs (IWGIA) Copenhagen.* Retrieved 3 November, from https://www.iwgia.org/en/news-alerts/news-covid-19/3553-statement-on-covid-19-and-indigenous-peoples-by-victoria-tauli-corpuz,-un-special-rapporteur-on-the-rights-of-indigenous-peoples.html

UN Human Rights (1998). Declaration on human rights defenders. Retrieved 3 January 2021, from https://www.ohchr.org/en/issues/srhrdefenders/pages/declaration.aspx

United Nations (2007). United Nations declaration on the rights of indigenous people. Retrieved 3 January 2021, from https://www.un.org/development/desa/indigenouspeoples/declaration-on-the-rights-of-indigenous-peoples.html

United Nations (2018). Global compact for safe, orderly and regular migration: Intergovernmentally negotiated and agreed outcomes. July 13. Retrieved 21 December 2020, from https://refugeesmigrants.un.org/sites/default/files/180713_agreed_outcome_global_compact_for_migration.pdf

Preface

This book has been developed during extraordinary times. The context for the development of the "socialising tourism" concept was the global crisis caused by the COVID-19 pandemic. It was the observation that injustice, inequity and an almost ecocidal assault on the natural environment gave rise not only to the pandemic but also to the way that many governments failed to act in the interests of the *whole* community and that the worst impacts were borne by the most vulnerable, marginalised and impoverished. Words and phrases entered the lexicon such as "vaccine privilege", "vaccine tourism", the "China flu" and "lockdown fatigue" that pointed to the sharp edges of class, privilege, racism, authoritarianism, resistance and unequal impacts that the pandemic illuminated in sharp relief. The disruptions of the pandemic on movement and travel have not only highlighted many of the injustices and inequities embedded within the phenomenon of tourism, but importantly have forced many of us to question and rethink tourism and our understandings of it entirely.

The article that inspired this work, "Socialising tourism for social and ecological justice after COVID-19", was published in *Tourism Geographies* in April 2020. We made an informal call for contributions to this edited volume in July 2020. The book proposal underwent peer review and was accepted in October 2020 by Routledge. Our authors were asked to submit their draft chapters by the end of October 2020. We committed to undertake rounds of editing and revision during the typically academic holiday times at Christmas and New Year with a deadline for manuscript submission in February 2021. We are grateful to C. Michael Hall as the commissioning editor for the Geographies of Leisure and Tourism Series, editor Lydia Kessell and all the dedicated staff of Routledge who helped this book come into being. We are also grateful to RL Starr (Wakayama University) for her support in indexing the book.

Our authors come from all around the globe and had different experiences and circumstances in the pandemic. Many experienced lockdowns, some had to pivot to online teaching under extreme time pressures, some had multiple caring roles, others experienced job precarity as universities dealt with difficult economic environments and others were impacted by the stark decline in the tourism industry, which was arguably the most negatively impacted by the global pandemic. In the midst of this all, we were asking them to draft chapters for this work with tight timelines and demands on their time for revisions and dialogue. We owe a debt of gratitude for their commitment and good humour that exceeds the usual circumstances that all such edited volumes entail. Not only did our authors deliver, and mostly on time, but they delivered extraordinary and impactful work. We commend each and every chapter in this book for your reading and each is sure to challenge your thinking, incite your emotions or trigger your commitment to ongoing action to create better futures in and through tourism.

We, the editors, were also not left unscathed by the pandemic and its effects. We will keep these stories private, but these experiences did shape our approach to the work. The duality between despair and hope was usually evident in our writing, in our conversations and in our visions of the future. We often said during the work to manifest this book that it was a joy and a distraction from the concerns, trials and fears we experienced during the pandemic backdrop to the project. We imagine you, our readers, will also be subject to this duality of fear and hope. We invite you on this journey with us of socialising tourism so that we may sway the balance away from fear and more toward hope. Hope and action are the catalysts to a more fair, just and sustainable future. The project of socialising tourism may be one important aspect to achieve this and through this book we begin.

Freya Higgins-Desbiolles, Adam Doering and Bobbie Chew Bigby

Introduction

Socialising tourism: reimagining tourism's purpose

Freya Higgins-Desbiolles, Adam Doering, and Bobbie Chew Bigby

Introduction

Tourism is a controversial and contested phenomenon. This starts at the most basic level when many people engaging in voluntary travel designate themselves as travellers, thus attempting to distance themselves from maligned tourists. Long gone are the days of the Grand Tour when touring was seen as a journey for cultivation and refinement of one's educational and cultural learning. Too often now tourists are viewed as hedonistic pleasure seekers who care little for anything beyond their own enjoyment on holidays. Simultaneously, the industry of tourism has come in for criticism for overwhelming some destinations through overtourism (Dodds & Butler, 2019; Goodwin, 2017), failure to deliver the promised benefits (Chalip & Costa, 2012) and being subject to a practice of "bugger it up and pass it down" (Wheeller, 1993, p. 125). Tensions, protests and resistance have grown in places as varied as Barcelona, the Galapagos Islands, Venice, Byron Bay, Kyoto and Mount Kilimanjaro. Something has gone awry in the evolution of tourism and we suggest here that thought should be given to the ways in which we might "socialise tourism" to set it on a better trajectory.

The term "socialising tourism" is a provocation for critical engagement. In recent political campaigns in countries such as the United States, the United Kingdom and Australia, the labels socialists and socialism were used as pejoratives to ensure one's opponents were dismissed as either utopians or Stalinist authoritarians. However, this term is richer and more complex than this limiting political label. First, tourism is a socialising activity in the most basic meaning: "to spend time when you are not working with friends or with other people in order to enjoy yourself" (Cambridge Dictionary Online, n.d.). However, there is another important connotation to socialising: "to train people or animals to behave in a way that others in the group think is suitable" (Cambridge Dictionary Online, n.d.). The term "socialised" might also mean: "provided or paid for by the government" (Cambridge Dictionary Online, n.d.). As these few initial forays demonstrate, this term is

DOI: 10.4324/9781003164616-14

rich with possibilities and invites different thinking about tourism and how we might engage with it.

This is not a sociology of tourism, although this edited work will intersect with sociological concerns in some discussions. Sociology is "the scientific study of human life, social groups, whole societies and the human world as such" (Giddens, 2009, p. 6). As many tourism analyses indicate, tourism is an important and "increasingly widespread social activity" shaping our world and may even offer "a lens through which people and society can be studied" (Sharpley, 2018, p. 20). Sociologists Cohen and Kennedy have asserted that tourism has helped shape globalisation with an "outreach greater than other powerful globalising forces" (2000, p. 213). Clearly tourism matters and for more than just employment, foreign exchange and economic growth.

Socialising tourism can be viewed as a revival of earlier thinking on "tourism as a social force" (Higgins-Desbiolles, 2006). Conceptualisations of tourism as a social force represented a pushback against the growing hegemony of neoliberalism and the power of the market to transform and limit tourism to its business aspects and claim it as an industry. This pervasive view of tourism as an industrial sector contributing to growth in economies has had significant repercussions on not only increasing the unsustainability of tourism but also diminishment of tourism's social possibilities. The view has become so pervasive in both the tourism academy and in the wider society that to think critically on tourism and to challenge its injustices may be interpreted to be anti-industry or even waging a "war on tourism" (see Butcher, 2020; Higgins-Desbiolles, 2021).

However, COVID-19 offered a circuit breaker to this ideological domination of the "tourism as industry" point of view. As Higgins-Desbiolles argued:

> The COVID-19 pandemic crisis has challenged the premises of neoliberalism that smaller government, individualism, and marketisation benefit people and society. Forms of government interventions, the redevelopment of social safety nets, and the significance of social caring and networks have been the primary responses to challenges of this crisis.
>
> (2020, p. 8)

Importantly, as a response to the extraordinary challenges of the COVID-19 crisis, neoliberal governments willingly adopted "socialised" policies in their responses to the pandemic. For instance, there have been large government expenditures to support businesses and employees severely impacted by the lockdowns (particularly in hospitality and tourism) and control measures the crisis necessitated. Such governments have also been forced to fund health, social and educational support packages (sectors they had previously white-anted through privatisation) in an effort to avoid widespread social unrest. In some cases, governments have decided to support the tourism industry, tourism jobs and tourism-dependent communities with temporary subsidies for citizen's holidays through holiday vouchers and other mechanisms.

Taking this historical moment as an opportunity to rethink tourism, this book explores the possibilities of socialising tourism for better outcomes. To this end, this introduction addresses these opening questions:

- What might socialising tourism mean?
- Why does tourism need to be "socialised"?
- How might tourism be socialised?
- What might we ask of tourism?

A first foray into socialising the stranger: First Nations

In setting such an agenda, the knowledges of Indigenous and First Nations peoples around the world offer rich insights into the philosophies, practices and spiritualities that might underpin such a socialisation agenda. We might turn to First Nations first as Indigenous and First Nations peoples have enduring hospitality, socialisation protocols and ceremonies over millennia that might offer exemplary insights into possibilities for socialising tourism. For instance, the Māori of Aotearoa/New Zealand have protocols and ceremonies for receiving and socialising visitors on the *marae*, the meeting ground of Māori *iwi* (tribes). Harvey (2003) explained these protocols as shaping guesthood (in his analysis of decolonising research methodologies), which are based on recognising local sovereignty arising from the host's marae serving as *turangawaewae*, the "standing place" of the host. Marae protocols of welcome, greeting and exchanges are protocols of "guest-making" as strangers are transformed into guests (Harvey, 2003, p. 134). But the foundation of this interaction and relationship is respect for the local people's authority as the sovereign peoples of that place. This is an excellent example of socialising the visitor.

In Australia, many Aboriginal First Nations peoples respect and follow their Dreaming laws, which in part explain how physical landscapes serve as reminders of protocols and the consequences for violating them. Aboriginal tour operator Quentin Agius shared a story that communicates one such Dreaming narrative that addresses proper protocols for entering another's Country:

> We talk about how Nookina came from the northern part of the country and came into southern part of country without permission, and then Nookina and Windera got into a fight.

> During that fight Windera got injured and he laid down in a certain special area, and where he laid down he became a part of country and you can see to this day the different coloured rocks that are parts of his body.
>
> (Clarke, 2015)

The very idea of hospitality is welcoming the stranger and questions on how the Other is received (see Scott & Hall, 2012). But this Dreaming narrative indicates that there are mutual obligations when visiting the home of another. The socialising values that might be gleaned from this foray into First Nations' practices and protocols include respect, relationships, reciprocity and responsibilities, and these are mutual but differential on the parts of the hosts and the guests (see also Chapter 1).

Using a socialisation lens, we might understand settler-colonialism, colonialism and other forms of dispossession of First Nations peoples as the most violent and damaging form of violation of host-guest protocols. As a result, the modern meeting grounds between Indigenous hosts and non-Indigenous tourists too often symbolises a good deal of what is wrong in marketised forms of tourism:

> It is a truism that to visitors to a new land – certainly in the case of early settlers – the original inhabitants were profoundly Other... In settler societies such as Canada, New Zealand and Australia, tourism development is often controlled by non-indigenous peoples and dominated by power structures that have originated through colonialism.
>
> (Amoamo & Thompson, 2010, p. 37)

This observation by two Māori First Nations scholars alerts us to remember and account for the origins of tourism in colonialism and imperialism, interwoven with forms of capitalism that emerged from these forces, because these still reverberate in current dynamics of tourism today. Additionally, First Nations' protocols provide a word of caution on proceeding with the introduction and imposition of tourism without proper socialisation into the appropriate ways to visit another people and their lands.

What might socialising tourism mean?

The socialisation of tourism offers one pathway to transform relations in tourism so that justice, equity and sustainability may be better secured. As explained earlier, the word "socialise" could hold multiple meanings, including following the principles of socialism; to act socially well in interactions with others; or to guide on proper ways of behaving with regards to society. In his discussion of "socialising the stranger", Scott provided an insight into the way socialising tourism could be understood. Scott explained: "Hospitality becomes an initiation of the process that would result in the socialisation and thus integration into the 'local' society" (2006, p. 57). Drawing on Scott, Higgins-Desbiolles proposed the concept of socialising tourism meant "[…] to make tourism responsive and answerable to the society in which it occurs" (2020, p. 617). She insisted that it is both tourists and tourism businesses that must be socialised into respecting the lifeways of the local community (often called "hosts") and serving the needs and interests of the local societies in which the tourists tour or the tourism industry offers tourism services.

Tourism can be a means for socialisation of tourists into the worlds of others and foster understanding and empathy for those struggling with social inequalities. As Parrish (2014) argued, sports and leisure socialised the young Ernesto Che Guevara through interactions with the oppressed and the poor, and this helped shape his concerns with social justice and his revolutionary trajectory. Through this example, we can intuit that socialisation of tourism must hold a concern with Others and communal bonds. It should present a stark contrast to the individualistic, hedonistic, self-focused and accumulative forms of tourism that have evolved from commercial tourism fostered under neoliberal market mechanisms.

Social tourism offers some useful insights into some of the particular ways by which tourism might be better socialised. As Minnaert et al. (2006) explained, there are two potential perspectives on social tourism: "visitor-related" and supply-side approaches to social tourism. The visitor-related forms of social tourism address the call of "tourism for all" by aiding those that are disadvantaged in any way to fulfil their desire to have a holiday (Minnaert et al., 2006, p. 8). There are a broad number of programs, activities and organisations that support such forms of social tourism, addressing many factors that inhibit people's enjoyment of holidays: low income, unemployment, aging, caring responsibilities, single-parent status, disabilities, etc. The supply-side view of social tourism focuses on forms of tourism that foster social interaction at the tourism destination. Seabrook (1995) explained:

> there is emerging a more convivial and interactive form of travel, a kind of social tourism; designed specifically to enhance and offer insight into the lives of people, which figures neither in the glossy brochures, nor in the media coverage of third-world countries.
>
> (cited in Minnaert et al., 2006, p. 8)

These forms of social tourism suggest that something important and valuable occurs through tourism experiences and encounters that have social value and should not be and cannot be left solely to the commercial tourism sector (see also Diekmann & McCabe, 2020). In terms of visitor-related social tourism, it is recognised that holidays support well-being and personal growth and that it is a matter of equity; that is, citizens should not be barred from such beneficial outcomes solely due to limited income or other barriers. Supply-side concerns with social tourism emphasise the human encounter made possible through tourism and seek to develop opportunities for that to be fostered rather than the commercial sector's emphasis on profits from tourist visitation. In their discussion, Minnaert et al. (2006) discussed the economics of both forms of social tourism; suggesting that visitor-related social tourism will require evidence that public monies supporting social tourism initiatives offer benefits in terms of reducing other social welfare costs; supply-side forms of social tourism represent a higher cost niche market that some individual tourists would be willing to pay. Recent research suggested that not only do disadvantaged individuals benefit from visitor-related social tourism opportunities through improved mental health, well-being and feelings of greater self-efficacy, but also the wider society might benefit in a number of ways (Kakoudakis et al., 2017). This research specifically showed that social tourism experiences can support job seeking behaviour as a result in the improvement in self-efficacy, and thus can make important positive contributions to both society and economy. Analyses of social tourism such as these succumb to such marketised and individualised views as a result of neoliberal instrumentalities that pressure advocates to justify social tourism spending on economic bases. From a socialising point of view, however, societies could prioritise social values over economic values (see Latouche, 2009) and visitors could be held more accountable to people and places (including but also beyond paying fair prices) for their holiday experiences when we remove neoliberal blinkers.

The contemporary, "Western" understanding of tourism comes from a rather narrow set of experiences and philosophies which results in emphasis on a highly individualistic and marketised tourism. In mainstream tourism literature, it can be difficult to find academic contributions to the critique of tourism that approach the topic from a "non-Western" perspective. One outstanding example is Inayatullah's "Rethinking tourism" (1995) which utilised, in addition to pacific and futures analysis, an Islamic perspective which was used to "deconstruct" tourism. Inayatullah claims an Islamic perspective centralises the phenomenon of pilgrimage and in particular the *hajj*, or pilgrimage to Mecca, which is one of the central pillars of Islam. Inayatullah describes it thus:

> Within … the Islamic world, all Muslims had to travel, they had to make the pilgrimage to Mecca. Indeed, travel or the accumulation of wisdom, *ilm*, was the essence of Islam. Travelling, visiting wise people, finding holy sites, was an integral part of life… the self travelled to gain spiritual knowledge… travelling, indeed was a microcosm of the spiritual journey of the Self.
>
> (1995, pp. 411–2)

While pilgrimage was not unique to the Islamic faith, what is perhaps striking is how central religious travel is to fulfilling obligations of the Islamic faith. Instead of the hedonistic focus of a great deal of contemporary, marketised tourism, this Islamic "tourism" is geared to spiritual growth and fostering of solidarity among the *ummah*, the community of believers within the Islam (see the autobiography of Malcolm X (1992) on this, Chapter 17 "Mecca"). Inayatullah charged: "the West … manufactures tourism services and the idea of tourism itself, which we have suggested is not a universal concept but a particular idea of a specific culture" (1995, p. 412). Inayatullah's contribution is valuable to any discussion of contrasting perspectives on tourism, because he reminds us that most tourism discourse emerges not only from the neoliberal economic paradigm but also from a narrowly "Western" set of experiences. Similarly, Hall (2006) has explored the role of Buddhist values in Asian tourism development and suggested these values work towards fostering a "middle way", the appropriate actions for each locality and activation of compassion (see Chapter 12 also on Buen Vivir).

It is also important to note that tourism has been used as a tool for political socialisation. The most obvious form is at the national level when tourism is utilised to fulfil nationalistic agendas and foster civic connections. For example, Zuo et al. (2016, p. 183) examined how tourism has been used as a tool of political education in China to "instil core political values and ideologies". Through a study of tourists' experiences of touring the Jinggangshan scenic area, a "red tourism" site, these authors explored the capacities and limits of tourism as a tool of political socialisation. Such forms of political socialisation through tourism were and are a feature of socialist and communist states (see Williams & Balaz, 2001). But political socialisation through tourism is more widespread than this and features in any nation where tourism is used for defining and building the bonds within the civic body (see, for instance, Doering & Kong, 2020). As Rasul Mowatt's chapter (Chapter 6) in this volume demonstrates, however, there are also very destructive forms of such political socialisation through tourism. In this chapter he reveals how Dylann Roof embarked on white nationalist road trips as preparation to carry out a massacre in South Carolina, USA, in 2015.

The most powerful instance of political socialisation through tourism is arguably the effort by American business leaders to capture the global economy for their profit-making and political agenda in the post-Second World War era of rebuilding in the war's aftermath and then the globalisation processes which followed based on globalised trade regimes. Patricia Goldstone's work *Making the world safe for tourism* (2001) is invaluable in explaining this. Describing these efforts as less about beneficent spreading of democracy and more about establishing neocolonialism for ongoing American profit, Goldstone wrote:

> […] the post-World War II travel offensive launched by Rockefeller's ally, American Express, […] sold tourism as an integral extension of the Marshall Plan. The resemblance of American Express' post war

> advertising campaign to speeches by James Robinson III's and other American Express executives during the approach of glasnost is not accidental, for American Express has appropriated democracy as its global brand; as the official company history proudly declares, "When dollar shortages are choking the arteries of international commerce, the American tourist plays a vital role in the economies of all free nations".
> (2001, p. 44)

Goldstone's work is vital to understanding the foundations of contemporary tourism and the reasons that it has been manipulated so easily for private profit and complicit in exploitation and human rights abuses. This sets a useful context for the argument of why tourism must be socialised differently for social and ecological justice.

Why does tourism need to be socialised for social and ecological justice?

The world is entering extraordinary times that are featured by enormous challenges. In a world nearing eight billion people, where many seek a quality of life that is based on a high consumption lifestyle, on a planet with finite resources, social and ecological pressures are mounting. We currently face pressures due to human impacts on the natural environment resulting in biodiversity loss, species extinctions, scarcities, pollution and whole habitats under threat. In recent years, extraordinary warnings have been issued by scientists that the world is on a dangerous pathway. For instance, in 2018 scientists explored the possibilities that human-induced climate change is leading us on a pathway to "hothouse earth": "If the threshold is crossed, the resulting trajectory would likely cause serious disruptions to ecosystems, society, and economies" (Steffen et al., 2018, p. 8252).

It is evident that tourism contributes to these concerns and even exacerbates these problems in ways that are becoming increasingly clear (see Scott et al., 2012). First is the sheer volume of tourism as evidenced through tourism statistics, including the 2019 data showing 1,459 million international tourism arrivals who generated US$1,487 billion in international tourism receipts according to the UN World Tourism Organisation (UNWTO) (UNWTO, 2019). This particular "tourism dashboard" fails to capture domestic tourism statistics, but it is certainly larger in terms of numbers of people travelling.

To support such massive movements of people and to cater to their tourism demand, enormous impacts occur. Studies of the impacts of tourism must be one of the most voluminous facets of tourism studies (e.g. Hall, 2008; Mason, 2003). The studies that document and explain the many types of negative tourism impacts – including social, environmental, economic, cultural, political and spiritual – provide insight into the specifics of the problems (Hall, 2008). In fact, some forms of tourism might be characterised as anti-social in the ways they have sometimes brought offensive, anti-social and abusive behaviour into the communities where these tourists holiday; this includes sex tourism, rave tourism, stag party tourism and other particularly transgressive forms of tourism. Certain forms of tourism are ruining life for the local communities and even displacing them. For instance, Budapest has become a hotspot for party tourism that causes negative impacts while delivering little economic benefit (Schlagwein, 2020). However, we lack a macro-level view, which really prevents us from understanding just how much tourism contributes to the downward trajectory we are on in terms of human impacts on natural environments, global environment change and exacerbation of social tensions (see Hall, 2008).

There are many aspects to these issues, but the one we will highlight here is the way communities around the world have been pressed to accept corporate forms of tourism development. Higgins-Desbiolles argued:

> Worse still is how communities seeking development are pushed into a tourism-dependent economy in their attempts to try to garner some opportunities for themselves in a global trading system geared to their continued under-development. In the process, they serve up their people to be the docile workforce so that tourists can enjoy inexpensive holidays in these imposed tourism playgrounds and tourism multinationals can extract wealth as a result.
>
> (2018, p. 158)

Tourism under neoliberal globalisation undermines the power of society to manage, control and benefit from tourism businesses operating in their communities because the global market they are tapping into runs outside of their control. This is a key catalyst to the recently documented problem of overtourism. Goodwin characterised overtourism as occurring when "hosts or guests, locals or visitors, feel there are too many visitors and that the quality of life in the area or the quality of the experience has deteriorated unacceptably" (2017, p. 1). The causes of overtourism varied according to the destination. The disruptive agents of the sharing economy, like Airbnb, were blamed for bringing more tourists into the heart of communities instead of just tourist sites. Cheap travel and package holidays enabled more people to take short city breaks and cruises, particularly in Europe. Social media played a role in popularising less-visited places, which went from being off-the-grid to "must-see" destinations overnight. The shifting focus of governmental tourism agencies saw them become almost exclusively marketing-focused with a singular goal of growing tourism (Higgins-Desbiolles, 2018).

Understandings of overtourism should be situated in the wider context of tourism development being fostered by the capitalist economy system for profit accumulation of multinational corporations and the global elite (Fletcher, 2011; Higgins-Desbiolles, 2008, 2018). As Fletcher argued:

> A small number of increasingly interrelated transnational tourism operators control much of the goods and services that tourists consume

> globally. In this respect, tourism expansion can be viewed as an instance of "accumulation through dispossession" that Harvey (2005) finds characteristic of neoliberal capitalism in general. These operators also control much of the advertising by which tourists are enticed to consume the products offered. Transnational tourism operators work hand-in-hand with other important tourism promoters, including international development agencies and national governments.
>
> (2011, p. 455)

Through this political economy lens, we can see how overtourism occurs through the pressures of multinational tourism corporations and affiliated others, who press for pro-growth approaches to tourism development. They lack concern for the limits of carrying capacity that a particular destination might be subject to, and in current neoliberal contexts of deregulation are not compelled to respect such limits. Another key aspect of this is the usurpation and privatisation of the commons which is a key feature of neoliberal capitalistic tourism (see Fletcher, 2016) and one significant source of the serious ecological injustices of tourism.

In their consideration of degrowth as a pathway to address such issues, Higgins-Desbiolles et al. (2019) proposed defining tourism by the rights, interests and benefits of the local community and thereby reorient the phenomenon entirely (more on this below). The justification for such a radical proposal can be seen, for instance, in the results of tourism surveys which indicate the ways in which local people feel alienated by tourism. As an example, a 2018 survey by the Hawaii Tourism Authority indicated that two-thirds of respondents agreed with the statement that "This island is being run for tourists at the expense of local people" (Hawaii Tourism Authority, 2019, p. 22).

While COVID-19 has temporarily caused the issue of overtourism to retreat from focus, we can anticipate it will return and will raise possibilities for tension, hostility and even violence, as has been witnessed in places such as Barcelona. Andrews has demonstrated that "violence is manifest in many aspects of tourism practices and encounters..." (2014, p. 5). Additionally, analyses have demonstrated that inappropriate tourism development can result in a violent deterioration in the quality of life for resident communities (see da Cal Seixas et al., 2014).

This brings us to considerations of the right of local communities to say "no" to inappropriate tourism developments, which has been too little discussed in tourism studies (see Robinson, 1999, as one exception). We would argue that the ultimate way to determine if tourism is properly socialised is when the local community has the capacity to say "no" to tourism and/or tourists and to deny tourists entry into their place (whether temporarily or more long term). This has occurred in places around the world in the context of the COVID-19 pandemic. As people sought to escape urban areas with their lockdowns and stresses, nearby holiday hotspots became concerned these city populations would spread the contagion to their communities and threaten to overwhelm their limited medical facilities. Indigenous communities were among such communities. The BBC reported that First Nations are "uninviting visitors", noting in British Columbia:

> In an effort to keep out outsiders, Haida Gwaii and the Central and North Coast, including the Heiltsuk and Kitasoo / Xai'xais Nations, have set up a coalition. They are working collectively to let visitors know that while they are valued and wanted, now is not the time to visit. The communities are simply too vulnerable to risk any loss.
>
> (Selkirk, 2020)

In another instance, on 3 August 2020, 30 members of Mutitjulu Aboriginal community blocked tourists' access to Uluru-Kata Tjuta National Park in central Australia because of fears they arrived from interstate COVID hotspots and might bring the disease into the vulnerable local Aboriginal community (Barnsley, 2020). Such cases demonstrate socialisation of tourism to the lesson that tourist right of access cannot override the community's right to ensure community benefit and well-being.

How might tourism be socialised for social and ecological justice?

Possibilities for socialising tourism include focusing on how we can socialise tourists for respectful and responsible behaviour during their holidays; how we might ensure the benefits of tourism are more equitably distributed within society; how we might socialise tourism industry businesses to assume appropriate roles and practices in the jurisdictions where they operate; how we might socialise governments for shaping tourism to the needs of the local communities where tourism occurs; and socialising our values to ensure that tourism is harnessed for social and ecological well-being. Indeed, the seeds for socialising tourism are already with us and we will provide some examples of these below. However, this section will also flag that there is much more to be done and some of the chapters in this volume will start this important work.

Attempts to socialise tourists through codes and protocols has had a lengthy history. These codes have been viewed as an important means to make tourism more ethical and responsible (i.e. Fennell & Malloy, 2007; Lovelock & Lovelock, 2013). An example of this is the International Ecotourism Society's code of conduct for ecotourism and ecotourists. Recently, Aotearoa/New Zealand in 2019 instituted a campaign called the Tiaki Promise (Tiaki, n.d.), which urged international tourists to assume the role of custodians of New Zealand and take care of it during their holidays (*tiaki* is a Maori word for care). Additionally, the north Pacific state of Palau developed the Palau Pledge (n.d.), which was marked by an official stamp in visitors' passports, asking tourists to be careful of the people and place of Palau during their visit.

There are also codes which have sought to reach other stakeholders in addition to the tourist. A good example of that is the UNWTO's "Global Code of Ethics for Tourism" (1999). This extensive document has sections addressed to all major stakeholders in tourism, including the tourists, the tourism industry, host communities and governments at all levels. However, as Castañeda (2012) explained, this document does not effectively promote greater equity and sustainability in tourism. Instead, the Code validates "laissez faire neoliberal expansion of tourism development" and "... unequivocally asserts the subordination of the heritage rights of destination communities to those of tourists through the use of its awkward yet very precise language" (2012, p. 49).

To ensure greater equity and social justice, there are well established and sophisticated programs and policies for social tourism which exist all around the world, as previously mentioned (see also Chapter 13). Social tourism programs and facilities are well known in parts of Europe such as Spain and Belgium. Brazil offers another example to consider with both federal state authorities and third sector actors contributing, as reported by de Almeida (2011). His brief analysis of one programme explained:

> SESC does not only strive to reduce the price of the holidays and the hotel rates for its members, but above all it aims to transform people by "[...] developing their intellectual and physical skills, knowledge and social interaction".
>
> (de Almeida, 2011, p. 488)

The example of Brazil models possibilities for combining hospitality, leisure, recreation and tourism for the full participation and well-being of the society. However, as studies show, it is the resourcing and dedicated implementation that are essential for success.

In thinking through how we might better socialise tourism, we might give consideration to how we can socialise the tourism industry. There are specific codes of conduct and accreditation schemes which are purported to guide businesses to ethical and sustainable conduct. These have been criticised as often voluntary and weak, with strongest attention to measures that support the business' benefit such as energy savings (see Hall, 2008; Mason, 2007). Similarly, the United Nations Sustainable Development Goals has been promoted as a pathway to greater sustainability but still works to sustain growth in production and consumption (see Boluk et al., 2019). More promising in terms of socialising tourism businesses is implementing the concept of business "social license to operate" (SLO) (see Williams et al., 2007). SLO has been better developed for the more well-recognised extractive industries and describes business efforts to obtain legitimacy and acceptance of their operations, even in cases when the impacts are negative.

As indicated earlier, tourism under neoliberalism can be as extractive and damaging as logging or mining and is often not the benign industry that some suggest (see Fletcher, 2011). Better developing the concept of SLO in tourism and analysing how to effectively implement it might be a key pathway to socialising tourism.

In order to achieve the socialisation of tourism for social and ecological justice, an essential focus must be to effectively address the role of governments. As already argued, under neoliberal conditions, too often governments support the interest of corporations at the expense of society and push an unsustainable growth ideology (Fletcher, 2011). Some of the positive outcomes of the COVID-19 crisis have been the demonstration that governments must support societal thriving first and also that social solidarity is the key to positive futures. This is how, for instance, tourist hotels were quickly commissioned to house the people experiencing homelessness during the crisis management phase. As a result of the crises of capitalism and capitalistic tourism, there are numerous considerations of alternative governance regimes and alternative economies which are envisioning new approaches (i.e. Hall, 2018; Higgins-Desbiolles, 2008; see also Chapters 5 and 13).

So far, this section has covered what already exists, which is clearly not yet sufficient to reorient tourism for securing social and ecological justice. The proposal by Higgins-Desbiolles et al. (2019) to redefine tourism by the local community has some potential for better socialising tourism. These authors explained:

> Tourism for sustainability and degrowth must focus on the needs and interests of the local community; what tourism industry interests have usurped for themselves under the label of the "host community". A redefined tourism could be described as: the process of local communities inviting, receiving and hosting visitors in their local community, for limited time durations, with the intention of receiving benefits from such actions. Such forms of tourism may be facilitated by businesses operating to commercial imperatives or may be facilitated by non-profit organisations. But in this restructure of tourism, tourism operators would be allowed access to the local community's assets only under their authorisation and stewardship.
>
> (2019, p. 1936)

This rethinking resulted in Figure 0.1, which illustrates the transformed relationships with such a definition. In this conceptualisation, the local community where tourism occurs is placed at the centre and this resets relationships with other key actors in tourism. This includes the tourist being socialised to be a guest rather than a demanding customer; governments recognising the authority of local community over tourism; and tourism businesses being socialised to earn and maintain their SLO. A justification for such an approach might be evident from our earlier discussion of First

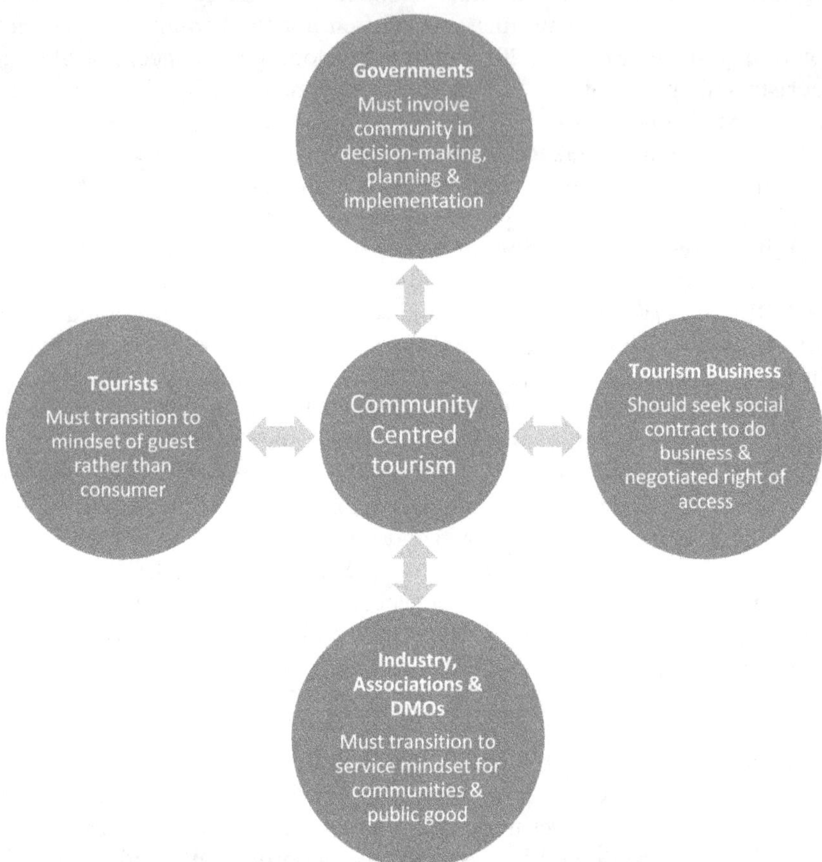

Figure 0.1 Community-centred tourism framework. Figure adapted with permission from Higgins-Desbiolles et al. (2019, p. 1937). Taylor and Francis Inc. http://tandfonline.com.

Nations protocols on Māori marae, when we explained the recognition of local sovereignty arising from the host's marae serving as turangawaewae, the "standing place" of the host. The place where tourism occurs is not a tourism destination; it is the local community's home, their standing place, and a place of uncompromisable value.

Finally, we must briefly turn to the values that are essential for this agenda of socialising tourism. In the discussions of First Nations pedagogy, we have already gleaned values of respect, relationships, responsibility and reciprocity. A socialising approach is a social approach that is based on relationality and must, by definition, be Other-oriented. However, this is not sufficient in a global order that is causing such large-scale damage and destruction. Socialising tourism must foster greater understanding of our interdependency as humans – our bonds, our need for each other as well as the environment that nourishes us. Feminist scholar Judith Butler (2020) has penned a recent treatise on interdependency, demonstrating that we are not isolated individuals but interdependent social beings with global obligations. This is a philosophy for socialising tourism as a global agenda.

What might we ask of tourism?

The project of "rethinking tourism" has been under development for several decades. This includes the important milestone when Deborah McLaren offered her book *Rethinking tourism and ecotravel* (1998) and Inayatullah's "Rethinking tourism" (1995) offering an Islamic perspective referred to earlier. As Higgins-Desbiolles (2006) asserted more than a decade ago, considerations of tourism as a social force have been overshadowed with the rise of neoliberal globalisation and the concomitant discourse and shaping of tourism as exclusively an industry limited to its commercial and business forms. But, as she demonstrated in that analysis, the older vision of "tourism as a social force" has powerful positive possibilities that are worth struggling for. With this introduction, we have extended this analysis with our considerations of socialising tourism. Here, we have explained what socialising tourism might mean, why such an effort is needed and ways we might go about achieving this agenda. This is only an exploratory overview and much more work remains to be done to more fully consider the possibilities.

In the wake of COVID-19, some tourism scholars noted how possibilities for transforming tourism emerged from the dynamics of the crisis (Lew et al., 2020). With this crisis, it became apparent to many that activities and services that are essential to public well-being, in particular healthcare, needed to be addressed as a universal social good rather than as a marketised commodity. As we close this initial analysis of the possibilities for socialising tourism, we might ask: could similar demands be placed on tourism? That is, could we shape tourism in such a way that it no longer causes widespread injustices and instead strives to serve as a universal social good? The chapters in this book will go some way to helping us answer such questions.

The organisation of this edited volume is divided into three sections.

Section 1. Socialising tourism as rethinking social relations

Socialising is the activity of engaging socially with others. We invited our authors to engage in critical questioning, including: How do we relate with one another and our environments with and through tourism? How are uneven and unequal social relations produced and maintained? How might they be reshaped? Socialising tourism requires a fundamental rethinking of the social relations and relationality of tourism at all scales: global, national, regional, local and individual. It is about how we relate with one another and how these social relations shape and are shaped by the political economy, geopolitics between nations, host-guest relations, colonial histories, race, gender and human-environment relations, to name a few. Examples such as toxic tourism that encourage affective and embodied engagement with polluted places to invoke change or decolonising tours used to rewrite tourism representations and politics in colonial settings are exemplary instances of making tourism work for the public good.

In Chapter 1, Andrew Peters and Simon Lambert offer Indigenous perspectives engaged with Māori hosting principles and Aboriginal Australia Welcome to Country/Acknowledgement of Country protocols. This analysis offers us insights into rethinking relations between tourists, culture and place. In Chapter 2, Bobbie Chew Bigby and Rebecca Jim analyse toxic tours as a tool for "environmental coalition building" through their case analysis of Tar Creek toxic tours. The tours offer the possibility of connecting visitors to places of environmental injustices while also allowing the hosts to use these tours for agency, empowerment and the search for justice. In Chapter 3, Sandro Carnicelli and Karla Boluk explore how we might move from the dominating practice of "carelessness" in contemporary tourism to a more "care-full" form of tourism. They emphasise the importance of "caring capacity" as a means to socialise tourism and propose a pedagogy and approach to foster such caring in tourists and others. Kokel Melubo and Adam Doering provide insights into the potential and constraints to involving local communities in tourism as tourists through a detailed analysis of experience in Tanzania's Northern Circuit in Chapter 4. From their analysis, it is made clear that a pandemic recovery process offers an opportunity to prioritise local communities as tourists but to do so requires overcoming colonial legacies and associated Western framings of tourism.

Section 2. Socialising tourism as rethinking ideology

Socialisation is the process of learning to behave in ways that are acceptable to society. The current socialisation into tourism's business-as-usual is no longer tenable. In contrast to recent calls to have tourism scholars to be "better aligned with industry", socialising tourism places critical theory, dissent, ideological critique, creativity, diversity and the forefronting of marginalised voices at the heart of tourism studies. Maintaining critical distance from corporate power and influence is essential if academic scholarship, education, NGOs and social activism are to offer critical and creative insights into how to do tourism differently. How have we been socialised to think about tourism? How has the COVID pandemic called these ideologies into question? How can we decolonise this hegemonic framing of tourism? What illusions and myths continue to obscure our insight into tourism's harsh realities?

In Chapter 5, Raoul Bianchi provides a political economy analysis of the global tourism system in order to challenge current conceptualisations of socialising tourism. He identifies the need for effective interventions against monopolistic corporate power, financial speculation and offshoring of capital, built on alliances between activists on these issues and workers exploited by these forces. Chapter 6 presents a different kind of challenge to the socialising tourism concept, as Rasul Mowatt recounts the white supremacist road trips undertaken by Dylann Roof before he carried out a massacre in 2015 at the Emanuel African Methodist Episcopal Church in Charleston, South Carolina, with the aim of starting a race war. Mowatt asks how can we embark on a project of socialising tourism when such unjust circuits and itineraries remain unchallenged by a committed agenda of truth-telling, reconciliation and restitution? Critical scholars Alana Dillette and Stefanie Benjamin provide Chapter 7 in which they present two narrative ethnographies exploring the ways they as two early career scholars have experienced ongoing colonialism in the tourism academy. Their work impresses the need to socialise the tourism academy and industry through a committed agenda of decolonisation. Their critical, feminist, decolonising work underscores how essential it is to address issues of power, privilege and associated oppressions performed in both contexts. Continuing in the American context, Kyle Kajihiro presents Chapter 8 where he examines Hawai'i's "DeTours" as an example of critical educational tours used as a tool to address ongoing colonisation and also the militourism that Hawai'i suffers. He argues these DeTours allow for decolonial place-making and weave webs of solidarity through the multiple and multilayer relationships that occur. These DeTours are not without their limitations and this is important to our socialising tourism agenda.

Section 3. Socialising tourism to build better collective futures

Socialising means adapting to social needs or uses and organising group participation to achieve these goals. Socialising tourism is therefore future directed and aimed at building better futures. These are not idyllic visions, but are futures grounded in what is currently occurring. Socialising tourism means engaging with the difficulties of the times and finding ways to fit tourism in the societies and ecologies in which it is occurring. For tourism to become more socially and ecologically just, it must find ways to better fit into local agricultural systems, local land uses, traditional ecological knowledge and residential policies and planning.

In Chapter 9, Shinji Yamashita considers forms of public tourism that can be identified as a response to the Great East Japan earthquake of 2011. He identifies these forms of public tourism as part of a new age of civic activities in Japan which have revealed the public good possibilities of tourism, particularly in a context of ongoing crises but also beyond. In Chapter 10, Adam Doering and Kumi Kato explain the search for "new light" in Fukushima, Japan, thereby illuminating possible alternative futures, moments of hope and bursts of beauty and creativity, even in the midst of devastation and destruction. Their work offers an affirmative, creative and exploratory ethos and methodology for scholars and practitioners of socialising tourism to consider. Carol Kline presents ideas of socialising tourism to ethical engagement with animals in Chapter 11. After documenting the forms of animal abuse and exploitation that occur in tourism, Kline engages with the emerging posthumanistic turn in tourism studies. She offers insights into species justice in tourism and how it can socialise tourism towards a more fulsome form of justice. In the penultimate chapter, Chapter 12, Natasha Chassagne and Phoebe Everingham explore the principles of Buen Vivir, which has emerged from the Latin American context. Buen vivir is about building economies based on concepts of well-being and represents one important paradigm challenging neoliberal, market fundamentalism and the growth ideology that accompanies it. The final chapter, Chapter 13, offered by Robert Fletcher, Asunción Blanco-Romero, Macià Blázquez-Salom, Ernest Cañada, Ivan Murray Mas and Filka Sekulova, offers one last critical challenge to the concept under study in this volume: socialising tourism. They critique the initial concept as introduced by Higgins-Desbiolles (2020), saying it focuses in on local community agency and action. In their view, this is not sufficient; they seek to activate social justice at scale across multiple levels. The three case studies they offer in this chapter support a vision they introduce of "eroding tourism" to arrive at a post-capitalist form of tourism that better benefits communities and ecologies. These, they argue, demonstrate the potential to combine diverse forms of action in different contexts and scales within an overarching strategy to erode capitalism and its sister, capitalistic tourism.

This text is a part of a long lineage of critical and engaged analysis and a contemporary expression of the dire need to rethink tourism as a social force. Because of current events and changes in tourism discourse, this rethinking of tourism has become increasingly urgent and possible. We are grateful to these authors for joining us in this project of fleshing out the provocative thinking on socialising tourism for social and ecological justice. These chapters take the reader on an engaging and stimulating journey in which we demand a lot of tourism.

References

Amoamo, M., & Thompson, A. (2010). (Re) Imaging Māori tourism: Representation and cultural hybridity in postcolonial New Zealand. *Tourist Studies, 10*(1), 35–55. https://doi.org/10.1177/1468797610390989

Andrews, H., ed. (2014). *Tourism and violence.* London: Routledge.

Barnsley, W. (2020, 3 August). Uluru national park closed after traditional owners block tourists amid coronavirus fears. Retrieved 5 August 2020, from https://7news.com.au/lifestyle/health-wellbeing/uluru-national-park-closed-after-traditional-owners-block-tourists-amid-coronavirus-fears-c-1212995.

Boluk, K., Cavaliere, C., & Higgins-Desbiolles, F. (2019). A critical framework for interrogating the United Nations Sustainable Development Goals 2030 Agenda in Tourism. *Journal of Sustainable Tourism, 27*(7), 847–64. https://doi.org/10.1080/09669582.2019.1619748

Butcher, J. (2020). The war on tourism. Spiked Online. Retrieved 5 May 2020, from https://www.spiked-online.com/2020/05/04/the-war-on-tourism/.

Butler, J. (2020). *The force of nonviolence: An ethico-political bind.* London: Verso.

Cambridge Dictionary (n.d.). Socialize. Retrieved 5 August 2020 from https://dictionary.cambridge.org/dictionary/english/socialize

Castañeda, Q. (2012). The neoliberal imperative of tourism: Rights and legitimization in the UNWTO global code of ethics for tourism. *Practicing Anthropology, 34*(3), 47–51.

Chalip, L., & Costa, C. A. (2012). Clashing worldviews: Sources of disappointment in rural hospitality and tourism development. *Hospitality and Society, 2*(1), 25–47. https://doi.org/10.1386/hosp.2.1.25_1

Clarke, A. (2015). Indigenous Australians tell you why you should take an Aboriginal tour. Retrieved 3 August 2020, from https://www.buzzfeed.com/allanclarke/why-your-next-holiday-should-be-an-indigenous-experience

Cohen, R., & Kennedy, P. (2000). *Global sociology.* Houndsmills: Macmillan Press.

da Cal Seixas, S. R., de Moraes Hoeffel, J. L., Botterill, D., Carnevale Vianna, P. V., & Renk, M. (2014). Violence, tourism, crime and the subjective: Opening new lines of research. In H. Andrews (Ed.), *Tourism and violence* (pp. 145–64). London: Routledge.

de Almeida, M. V. (2011). The development of social tourism in Brazil. *Current Issues in Tourism, 14*(5), 483–9. https://doi.org/10.1080/13683500.2011.568057

Diekmann, A., & McCabe, S. (Eds.). (2020). *The handbook of social tourism.* Cheltenham: Edward Elgar Publishing.

Dodds, R., & Butler, R. (Eds.). (2019). *Overtourism: Issues, realities and solutions.* Berlin: De Gruyter.

Doering, A., & Kong, T. (2020). Performative nationalism in Japan's inbound tourism television programmes: YOU, Sekai! (The World), and the tourism nation. In H. Endo (Ed.), *Understanding tourism mobilities in Japan* (pp. 138–57). London: Routledge.

Fennell, D., & Malloy, D. (2007). *Codes of ethics in tourism: Practice, theory, synthesis.* Clevedon: Channel View.

Fletcher, R. (2011). Sustaining tourism, sustaining capitalism? The tourism industry's role in global capitalist expansion. *Tourism Geographies, 13*(3), 443–61. https://doi.org/10.1080/14616688.2011.570372

Fletcher, R. (2016). Cannibal tours revisited: The political ecology of tourism. In M. Blàzquez, M. Mir-Gual, I. Murray, & G. X. Pons (Eds.), *Turismo y crisis, turismo colaborativo y ecoturismo.* XV Coloquio de Geografía del Turismo, el Ocio y la Recreación de la AGE. *Monografies de la Societat d'Història Natural de les Balears,* 18, *23,* 19–29. SHNBUIB-AGE.

Giddens, A. (2009). *Sociology* (6th ed.). New Delhi: Wiley.

Goldstone, P. (2001). *Making the world safe for tourism.* New Haven, CT: Yale University Press.

Goodwin, H. (2017). The challenge of overtourism. Responsible tourism partnership working paper 4. Retrieved 3 January 2020, from http://www.millennium-destinations.com/uploads/4/1/9/7/41979675/rtpwp4overtourism012017.pdf

Hall, C. M. (2006). Buddhism, tourism and the middle way. In D. J. Timothy & D. H. Olsen (Eds.), *Tourism, religion and spiritual journeys* (pp. 172–85). London: Routledge.

Hall, C. M. (2008). *Tourism planning: Policies, processes and relationships* (2nd ed.). Harlow: Pearson.

Hall, C. M. (2018). Anarchism and tourism: Coming sometime and maybe. *Tourism Recreation Research, 43*(2), 264–7. https://doi.org/10.1080/02508281.2018.1430717

Harvey, D. (2005). *A brief history of neoliberalism.* Oxford: Oxford University Press.

Harvey, G. (2003). Guesthood as ethical decolonizing research method. *Numen, 50*(2), 145–56.

Hawaii Tourism Authority (2019). HTA resident sentiment survey 2018 highlights. Retrieved 8 September 2020 from https://www.hawaiitourismauthority.org/media/2984/resident-sentiment-presentation-to-hta-board-01-31-2019.pdf.

Higgins-Desbiolles, F. (2006). More than an "industry": The forgotten power of tourism as a social force. *Tourism Management, 27*(6), 1192–208. https://doi.org/10.1016/j.tourman.2005.05.020

Higgins-Desbiolles, F. (2008). Justice tourism and alternative globalisation. *Journal of Sustainable Tourism, 16*(3), pp. 345–64. https://doi.org/10.1080/09669580802154132

Higgins-Desbiolles, F. (2018). Sustainable tourism: Sustaining tourism or something more? *Tourism Management Perspectives*, 157–60. https://doi.org/10.1016/j.tmp.2017.11.017

Higgins-Desbiolles, F. (2020). Socialising tourism for social and ecological justice after Covid-19. *Tourism Geographies.* https://doi.org/10.1080/14616688.2020.1757748

Higgins-Desbiolles, F. (2021). The "war over tourism": Challenges to sustainable tourism in the tourism academy after COVID-19. *Journal of Sustainable Tourism, 29*(4), 551–69. https://doi.org/10.1080/09669582.2020.1803334

Higgins-Desbiolles, F., Carnicelli, S., Krolikowski, C., Wijesinghe, G., & Boluk, K. (2019). Degrowing tourism: Rethinking tourism. *Journal of Sustainable Tourism, 27*(12), 1926–44. https://doi.org/10.1080/09669582.2019.1601732

Inayatullah, S. (1995). Rethinking tourism: Unfamiliar histories and alternative futures. *Tourism Management, 16*(6), 411–5. https://doi.org/10.1016/0261-5177(95)00048-S

Kakoudakis, K. I., McCabe, S., & Story, V. (2017). Social tourism and self-efficacy: Exploring links between tourism participation, job-seeking and unemployment. *Annals of Tourism Research, 65*, 108–21. https://doi.org/10.1016/j.annals.2017.05.005

Latouche, S. (2009). *Farewell to growth.* Oxford: Polity Press.

Lew, A. A., Cheer, J. C., Haywood, M., Brouder, P., & Salazar, N. B. (2020). Visions of travel and tourism after the global COVID-19 transformation of 2020. *Tourism Geographies, 22*(3), 455–66. https://doi.org/10.1080/14616688.2020.1770326

Lovelock, B., & Lovelock, K. (2013). *The ethics of tourism.* London: Routledge.

Malcolm X (1992). *The autobiography of Malcolm X,* with Alex Haley and Ossie Davies. New York: Ballantine Books.

Mason, P. (2003). *Tourism impacts, planning and management.* Amsterdam: Butterworth- Heinemann.

Mason, P. (2007). "No better than a band-aid for a bullet wound!" The effectiveness of tourism codes of conduct. In R. Black & A. Crabtree (Eds.), *Quality assurance and certification in ecotourism* (pp. 46–64). Wallingford: CABI.

McLaren, D. (1998). *Rethinking tourism and ecotravel* (1st ed.). Bloomfield, CT: Kumarian.

Minnaert, L., Maitland, R., & Miller, G. (2006). Social tourism and its ethical foundations. *Tourism, Culture and Communication, 7*(12), 7–17.

Palau Pledge (n.d.). Retrieved 20 October 2020, from https://palaupledge.com/

Parrish, C. (2014). Building character and socialising a revolutionary: Sport and leisure in the life of Ernesto 'Che' Guevara. *The International Journal of the History of Sport, 31*(7), 747–59. https://doi.org/10.1080/09523367.2014.901757

Robinson, M. (1999). Collaboration and cultural consent: Refocusing sustainable tourism. *Journal of Sustainable Tourism,* 7(3–4), 379–97. https://doi.org/10.1080/09669589908667345

Schlagwein, F. (2020, 24 July). Budapest moves to make party tourism a thing of the past. *DW Online.* Retrieved 6 August 2020, from https://www.dw.com/en/budapest-moves-to-make-party-tourism-a-thing-of-the-past/a-54307202

Scott, D., Hall, C. M., & Gossling, S. (2012). *Tourism and climate change: Impacts, adaptation and mitigation.* Abingdon: Routledge.

Scott, D., & Hall, S. (2012). Hospitality's others: A conversation. In S. Tallant & P. Domela (Eds.), *The unexpected guest. Art, writing and thinking on hospitality* (pp. 291–8). London: Art Books and Liverpool Biennial of Contemporary Art.

Scott, D. G. (2006, March). Socialising the stranger: Hospitality as a relational reality (Dissertation). Retrieved 3 April 2020, from http://hdl.handle.net/10523/1283

Seabrook, J. (1995). Far horizons. *New Statesman and Society, 8,* 22.

Selkirk, D. (2020, 26 May). Why First Nations communities are uninviting visitors. *BBC Online.* Retrieved 5 October 2020, from http://www.bbc.com/travel/story/20200525-why-first-nations-communities-are-uninviting-visitors

Sharpley, R. (2018). *Tourism, tourists and society* (5th ed.). London: Routledge.

Steffen, W., et al. (2018). Trajectories of the earth system in the anthropocene. *Proceedings of the National Academy of Sciences of the United States of America, 115*(33), 8252–9. https://doi.org/10.1073/pnas.1810141115

Tiaki: Care for New Zealand (n.d.). Retrieved 20 October 2020, from https://tiakinewzealand.com/.

UNWTO (1999). Global code of ethics for tourism. Retrieved 20 October 2020, from https://www.unwto.org/global-code-of-ethics-for-tourism.

UNWTO (2019). Global and regional tourism performance. Retrieved 28 October 2020, from https://www.unwto.org/global-and-regional-tourism-performance

Wheeller, B. (1993). Sustaining the ego. *Journal of Sustainable Tourism, 1*(2), 121–9, https://doi.org/10.1080/09669589309450710

Williams, A. M., & Balaz, V. (2001). From collective provision to commodification of tourism? *Annals of Tourism Research, 28*(1), 27–49. https://doi.org/10.1016/S0160-7383(00)00005-0

Williams, P., Gill, A., & Ponsford, I. (2007). Corporate social responsibility at tourism destinations: Towards a social license to operate. *Tourism Review International, 11,* 133–44.

Zuo, B., Huang, S., & Liu, L. (2016). Tourism as an agent of political socialisation. *International Journal of Tourism Research, 18*(2), 176–85. https://doi.org/10.1002/jtr.2044

Section I

Socialising tourism as rethinking social relations

1 "Wominjeka"/"haere mai"

The role of Indigenous ceremony in socialising tourism

Andrew Peters and Simon Lambert

Introduction

The global COVID pandemic that gripped the world beginning in 2020 has highlighted a number of problem areas for neoliberal societies the world over, including pollution levels, traffic congestion, waste management as well as mental health and domestic violence crises (Nelson, 2020; Zambrano-Monserrate et al., 2020). Increasingly, as neoliberal governments are called upon to act, they are facing ideological challenges in prioritising the needs of people and societies over those of market capitalism (Higgins-Desbiolles, 2020). In addition, positive impacts that include improved personal hygiene practices from social distancing (Zambrano-Monserrate et al., 2020) and a pivot from private enterprise to providing for social needs and causes (Ajifowoke, 2020) suggest the potential for a broad ideological shift at both micro (personal) and macro (society and industry) levels.

For Indigenous peoples, this shift is both welcome and unsurprising. For millennia, the fundamental principle of connection has underpinned the resilience and sustainability of Indigenous groups in all parts of the globe. In contemporary society, such notions are often challenged and contested as outdated and primitive and argued to have no part in the "modern world". However, there are certain human characteristics that are evident within Indigenous cultures that remain crucial to successful human interaction. Diplomacy is an ancient art that continues to be a powerful tool for nations to widen and deepen their influence (Beier, 2009). While modern diplomacy has seen some states co-opting Indigenous performances, arts and crafts to present a unique face to international partners, Indigenous communities continue to assert their collective rights and sovereignty as offering pathways to a more harmonious world through direct engagement.

This chapter will discuss the role of Indigenous ideologies in privileging and maintaining connections between not only human beings, but between all things. Using the concepts of "welcome to land (or country)" as a case study, we will explore the aspects of Indigenous Knowledge that are the foundation of connections to land and country and how they may play out in a socialising tourism future. Specifically, we will look at examples of

DOI: 10.4324/9781003164616-1

Māori cultural frameworks for welcoming and hosting visitors in Aotearoa New Zealand and Welcomes to and Acknowledgements of Country in Australia and how these protocols can establish and maintain a stronger sense of connection for tourists to place, culture and history. Such ceremonies are vital aspects of culture and cultural connection for Indigenous peoples the world over, as they provide ways to maintain and affirm significant practices within a predominantly colonial contemporary landscape. In addition, as we will discuss, they also provide clear, identifiable and connectable ways for non-Indigenous peoples to learn, understand and respect ancient cultures and practices.

Culture has remained a dominant factor in our (post)colonial world. Despite what can be called an onslaught of colonial, neoliberal factors that shape modernity and contemporary society, culture – particularly Indigenous cultures – stand strong and proud, and emerge as a key in socialising tourism and, in a broader sense, strengthening the fabric of society. This chapter looks at some specific examples of this neoliberal "onslaught" in and through tourism and the ways that Indigenous cultural resistance and response through the performance of ceremonies provide a unique answer.

Tourism growth, globalisation and Indigenous tourism

Tourism has long been a powerful force in bringing people, and thus difference, together. Like all cultures, tourism also has a history that has grown and evolved. The growth of tourism as a global sector mirrored the global capitalist growth from the 1970s, driven by the neoliberal agendas of economic growth pursued by countries such as the United States, the United Kingdom, Canada, Aotearoa New Zealand and Germany (Mosedale, 2016). During the 1980s, economic policy shifted from a Keynesian approach of involvement of the state (e.g. public-owned assets such as public transport and utilities) to more neoliberal philosophies of efficiency, lower taxes and lower debt designed to create a more indirect government role in the economy, thus creating an environment for high economic growth (Dekker, 2020; Mosedale, 2016).

Despite occasional shocks, tourism has seen phenomenal growth, with international tourist arrivals showing virtually uninterrupted growth for over 70 years: from 25 million in 1950 to 277 million in 1980, to 438 million in 1990, to 681 million in 2000, to 880 million in 2010 and to the most recent figure of 1.5 billion in 2019 (UNWTO, 2010, 2020). In addition, the industry has spread to now include destinations that were relatively unknown in the 1980s. While the growth and spread of tourism can be attributed to a number of factors, significantly the increase in globalisation and commercialism throughout the world has created "globalised" products that give consumers essentially the same experience wherever they are through processes that can operate and be reproduced on a global scale (Mowforth & Munt, 2015).

The proliferation of information technology, and specifically the internet, has created a much faster and wider spread of not only information, but people – including tourists.

Globalised tourism products such as accommodations, tours and theme parks therefore seek to appeal to a mass market, and it is often possible to replicate the visitor experience at each facility at any location. Examples include some of the large international hotel chains such as Hilton and Hyatt, organised visitor tours such as Contiki Tours (that are predominantly conducted in English) and attractions such as theme parks. Each of these examples will provide the same visitor experience whether travelling in Europe, North or South America, Asia, Africa or Australasia. In addition, visitor service is based on class (i.e. cost) rather than the visitor's origins or homeland. This "sameness" is a feature of globalisation; this is a major selling point to the modern mass tourist, thus making a major contribution to the large growth of global tourism. In this context, tourism can also be considered an example of colonial expansion, imposing methods of operation upon tourism producers and their products that follow the neoliberal objectives of capitalism that drive globalisation.

In addition, the neoliberal nature of mass tourism saw it subsume social tourism as the primary form of tourism throughout the world. In this context, "social tourism", although a rather nebulous term, refers to the 20th-century concept of "tourism for all" and the associated policies that enabled travel and tourism for more than the elite classes, providing more low-cost alternatives with broader appeal and access and standardisation of some products (see for example Higgins-Desbiolles, 2006; Minnaert, 2016; Segreto et al., 2009). The privatisation of key tourism assets such as transportation (air travel as well as public transport such as trains and buses) and health care gave rise to huge increases in tourists and travel patterns, and also contributed to the emergence of modern sectors such as medical tourism (Minnaert, 2016). This "colonisation" of tourism is particularly evident in the Asia Pacific region which has developed into a key destination for world travellers over the past two decades and is projected to grow in the future (UNWTO, 2015). Indeed, a large number of the emerging destination countries in the 21st century have been colonised at some point in their history. In the face of globalisation and the competitive nature of tourism as a business, culture and cultural difference assumes a role of great importance in product differentiation. Indigenous tourism has thus emerged as an important sector in creating a point of difference for tourists in this increasingly globalised world.

Emerging as both a legitimate business sector and as a legitimate field of academic inquiry in the early 1990s, Indigenous Tourism focuses on how Indigenous cultures, peoples and histories are utilised to attract visitors and create unique experiences. Additionally, the world's tourists are increasingly looking to learn about cultures and lifestyles different from their own, and are thus attracted to Indigenous tourism. As a tourism sector, Indigenous

Tourism provides remarkable opportunities to socialise tourism in the 21st century. As an emerging academic field in the 1980s and 1990s, it was subjected to a great deal of scrutiny around terminology and purpose, analysis that also reflected the growing global interest in Indigenous cultures more broadly. As such, phrases and semantics were given attention, including the term "Indigenous" and its applicability to tourism (Butler & Hinch, 2007; Ryan & Aicken, 2005). Words such as Aboriginal, Native, First Nations, Cultural and Ethnic were all used to describe tourism that focused on local, Indigenous cultures and peoples, and perhaps indicated a "confusion" created by a clear lack of broader knowledge and awareness of Indigenous cultures and histories in the early 1990s around the globe (Peters, 2017).

One of the more useful definitions in this field comes from Hinch and Butler, who suggested that "Indigenous tourism refers to activities in which Indigenous people are directly involved either through control and/or by having their culture serve as the essence of the attraction" (2007, p. 5). As such, the core principles of Indigenous cultures provide a unique and wonderful opportunity for tourism products that both satisfy an ever-increasing demand from tourists and enable a greater socialising imperative for future tourists.

Indigenous cultures and connection

There are broad aspects of Indigenous cultures that readily lend themselves to this potentially new direction that is more people-focused. For many thousands of years, Indigenous cultures the world over have maintained strong connections to their histories, lands and waters, cultures and languages in many ways. Evidence is laced through many contemporary societies, notably in the wake of the United Nations Declaration on the Rights of Indigenous Peoples (United Nations, 2007). Indigenous demands for a voice in all facets of their lives are acknowledged with, for example, Aboriginal flags flying in and around thousands of Australian buildings, Māori language programs in schools in Aotearoa New Zealand and Indigenous guardians and rangers employed in the conservation of ecosystems in Canada, to name but a few. Each exhibits an ongoing and continual cultural pride that is intergenerational and grounded in story. One of the fundamental aspects that allows, encourages, supports and promotes this cultural pride is the piecing together of past and present, stranger and family, animal and human, living and dead. For Indigenous peoples the world over, connection is a key to who we are. Our stories, whether they be ancient myth, past histories, contemporary experiences and future aspirations, act as the connective tissue between all things. "Story telling is the original classroom where history, morality and knowledge about people, places and the world are relayed to each new generation" (Langton, 2018, p. 34). Understanding our place in these stories is a crucial aspect of our own cultural identities. Storytelling is also emerging in academia as a vital process in maintaining and enhancing Indigenous concepts of mental and physical health, traditional (scientific) knowledges and restoring connections to kinship and Country (Bodkin-Andrews et al., 2016).

Our connections as Indigenous peoples are vital to our cultural survival and involve our connections to all things in our world, including land, language, history, people, animals and spirits. However, feeling a connection to land or a place is not necessarily an exclusively Indigenous characteristic. For example, many non-Indigenous people feel a deep sense of loss when their connection to "home" is taken away or destroyed by natural disasters such as earthquakes and bushfires. This sense of loss goes beyond the economic and emphasises the human aspect. As such, this sense of connection to land and country – and associated loss of land and country – should not be completely foreign to non-Indigenous peoples (Peters, 2017). This was emphasised in Australia recently with the devastating bushfires in late 2019.

Likewise, in Aotearoa New Zealand, Māori identity is implicitly and explicitly connected to the natural environment (Ataria et al., 2018). *Pākehā* (non-Māori) New Zealanders also express considerable emotional attachment to the environment and are increasingly supportive of Māori rights (*tino rangatiratanga*, sovereignty). Gaps in the research on Māori tourism noted by Zygadlo et al. (2003) have been narrowed through considerable work at both the tribal level and governmental level. In 2019, there were 234 Māori tourism businesses employing 11,100 people (Stats NZ, 2019). McIntosh and Zahra (2007) analysed the experiences of 12 Australian volunteers engaged by a Māori community in the North Island of New Zealand in 2005, finding that subsequent interactions and cultural experiences were mutually beneficial. Maintaining contemporary cultural connections remains a vitally important part of contemporary cultural identity.

Neoliberalism and disconnection

The political and economic growth over the past few decades, as discussed above, has brought with it a number of social and cultural problems. At the forefront is the increase in unequal distribution of wealth and income in most Western capitalist societies that, according to Kearney, "threatens social cohesion" (2017, p. 3). From a tourism point of view, this is exemplified in the growth of tourism after the Second World War to what became "mass tourism". During the 1960s in particular, tourism development adopted an approach that became known as "boosterism", where the supply-side of tourism was prioritised in order to increase visitor numbers, with very little regard for impacts other than economic (Hall, 2008).

Mass tourism became the focal point of development and tourism's exemplar of neoliberalism and globalisation. Development occurred in almost all parts of the capitalist (Western) world and was largely driven by goals of economic and political development. This included fast-paced developments designed to cater to thousands of tourists simultaneously, such as high-rise accommodation and large theme park attractions. Advances in information and transportation technology contributed to the conditions for mass tourism, with air travel and television providing greater desire and access (Hall, 2003). In Europe especially, changes in lifestyles, increased wealth and income, greater interest in "other places" and greater transportation freedoms all contributed to an 800% growth in travel from the 1960s to the 1990s (Lickorish & Jenkins, 1997).

As with most areas of society, however, this was also part of an intrinsic evolution – increased tourism brought increased focus from governments, industry and the academic world. The growth of mass tourism brought with it increased scrutiny, particularly the negative effects on local cultures and ecosystems and a "desire for more sustainable forms of tourism" (Minnaert, 2016, p. 132) that has also been labelled "new tourism". Sustainable tourism became a central focus of both tourism development and academic inquiry, as not only were the negative impacts addressed and mitigated, but economic opportunities were also recognised. Although in a tourism sense there is widespread subscription to a clear view of sustainability as outlined in the 1987 Brundtland Report, the discourse around "sustainability" and "sustainable tourism" is much more complex and nuanced. For example, Mowforth and Munt suggested the need for sustainability in areas such as politics, culture and lifestyle, in addition to profits (2015).

The discourses of this "new" tourism also highlighted the disconnection that had occurred between tourists and their own realities, tourists and hosts and in many cases host communities from themselves. Authenticity became an issue as many hosts began to alter their "normal" selves to suit the perceived wishes of the tourists, inspiring Dean MacCannell to develop the concept of "staged authenticity" (MacCannell, 1973), as tourists sought an authenticity that was different from their own reality (González Tirados, 2011).

This disconnection became a marker of neoliberalism in the modern world where, as Higgins-Desbiolles (2020) asserted, economic imperatives and market values have become embedded in social and welfare institutions that should be meeting the needs of society. We suggest here that a (re)connection to Indigenous values, cultures, histories and knowledges can play a key role in bringing these values and goals more in line with community expectations as part of the socialisation of tourism.

The role of Indigenous protocols

For Indigenous peoples today, culture is expressed in a myriad of ways. As mentioned above, storytelling is a crucial aspect of culture and cultural education for Indigenous groups. Knowledge is passed on through Elders and their stories in ways that affirm cultural identity and pride in those listening. Storytelling through Elders emphasises cultural and spiritual components of culture (Kunnie & Goduka, 2006). It is a vital tool for transferring knowledge and culture that comes in many forms. Our focus in this chapter is on how particular Indigenous protocols – namely Welcome To Country

(Australia) and the *Haka* and *Pōwhiri* (Aotearoa New Zealand) – are vital vehicles for knowledge sharing and provide a unique and invaluable opportunity to provide a connection for tourists.

Indigenous Peoples have highly attuned rituals and ceremonies reflecting their ancient concepts of how to engage with other peoples and societies. These practices deal with issues of connection and diplomacy, and not only retain their contemporary relevance, but have increasing value as future diplomatic efforts encompass diverse and self-conscious cultural groupings at both subnational and international scales – such as those seen in tourism. This chapter describes how Indigenous Peoples achieve new roles as diplomatic agents at the fluid borders of international exchange. In particular, the experiences of Māori in Aotearoa New Zealand will be presented, as they have increased their presence in international trade from a recent history of "performing" in arts and culture, sports, tourism and military roles. We suggest that in seeking wider relationships while promoting and protecting self-determined cultural logics, Indigenous Peoples have fundamental insights into how ethical exchanges can be articulated and maintained in host-tourist interactions.

In both cases, the performances of these cultural ceremonies serve to both affirm culture and its legitimacy in an often paradoxical, neoliberal world. These also allow non-Indigenous peoples to find ways to recognise, realise, respect and connect with ancient cultures that essentially always exist and operate "around" them – and always have. For many non-Indigenous peoples, especially those in countries without formal treaties such as Australia, this is a revelation. Indigenous culture is often assumed to be an adjunct to contemporary Western society as well as relegated to a precolonial past. However, cultural performance in a contemporary setting, such as a tourist attraction or destination area, allows a two-way reciprocal journey of learning about respect, caring and sharing – hallmarks of Indigenous cultures and peoples.

Australia – Welcomes and Acknowledgement of Country

In Australia, the Welcome to Country and Acknowledgement of Country ceremonies have become an important part of opening many events in almost all walks of life. Such ceremonies take place before events in all levels of government, schools, universities, community groups and sporting clubs. Although they have been subject to some criticism (see below), these ceremonies have assumed a role as an important marker of cultural recognition.

A formal Welcome to Country is only performed by a member (usually an Elder) of the Traditional Owners of the land on which it is performed. Welcome to Country ceremonies are usually longer and more involved, as they may also include song, dance, smoking ceremonies and sometimes exchanging of gifts. In each case, the gesture is a method of connecting with spirits and ancestors in order to pay respects to their own customs and traditions.

An Acknowledgement of Country is, as Emma Kowal asserts, a "twin ritual" to the Welcome to Country (Kowal, 2015). Acknowledgements are performed by non-Indigenous people or Indigenous people who are not Traditional Owners of the land on which it is performed. They are normally shorter in length and involve only the speech element. Acknowledgements are intended to acknowledge the Country as well as Elders and ancestors on which the event is taking place.

For the purposes of this chapter, it is not our aim to compare and contrast these two ceremonies, but rather bring them together to discuss the history, role and significance of each in establishing contemporary connections to land, culture and history. As such, future references to both will be as "welcome ceremonies".

The origins of the contemporary welcome ceremonies remain a little unclear. Scantlebury (2014) suggested that it may be traced back to the 1973 Aquarius Arts and Lifestyle Festival held in Nimbin, N.S.W. Scantlebury argued that this festival "was directly aligned with 'back to the land' philosophies" and was the first event in Australia to seek permission from Traditional Owners for use of the land (2014, p. 1). Others, however, point to a 1976 Welcome done by Richard Walley, Ernie Dingo and the Middar Theatre group for a group of Māori and Cook Island visitors as the first example of the "modern" Welcome to Country (Kowal, 2015; Penberthy, 2016) that "anthropologist Grant McCall thinks…was adapted from the Māori *powhiri*" (Kowal, 2010, p. 15). The only real consensus from most authors is that the specific origination of the contemporary welcome ceremonies is not clear (Kowal, 2010; Merlan, 2014; Pelizzon & Kennedy, 2019).

The purpose of these welcome ceremonies is a little easier to identify. At their very core, they acknowledge culture, history and place, and the convergence of these three aspects in a contemporary setting. In many cases (such as those performed by one of the authors), they also recognise and acknowledge non-Indigenous connections to this "trio of tradition". However, these ceremonies assume an important role in contemporary issues of race and culture. At face value, they act at the interface of Indigenous-non-Indigenous relations as examples of identity and cultural survival (Pelizzon & Kennedy, 2019) and cultural affirmation. The mere acknowledgement of Country in this way allows us to "immediately challenge hegemonic claims of land by colonial institutions" (Pelizzon & Kennedy, 2019, p. 25).

For some non-Indigenous people, these events are confusing, unnecessary and have even been labelled "divisive" and "racist" (Bolt, 2010) – often creating a perception (especially in Australia) that they "make intruders of non-Aboriginal people" who may not have ever lived anywhere else (Bolt, 2010). In addition to labelling them "racist", Andrew Bolt and Piers Ackerman have also called these ceremonies "inauthentic", a position that Bodkin-Andrews et al. (2016) designated as "inaccurate and essentialist". It is worth noting however that without context, the cultural elements of a welcome or acknowledgement can create appropriated and commodified practices that lack political meaning (Pelizzon & Kennedy, 2019, p. 24). While there has been healthy debate about the merits of these contemporary ceremonies (e.g. see Fredericks, 2013; Kowal, 2010, 2015; McKenna, 2014; Pelizzon & Kennedy, 2019), we prefer to focus on their ability to connect peoples, cultures and histories in contemporary settler-colonial settings.

Emerging from the precolonial traditions of connecting with spirits and land, these ceremonies highlight the *human* place in this context. Such ceremonies involve a recognition of the concept of "Country" for Aboriginal people. It is a multilayered web of relationships between all things. Country is more than a geopolitical object/construct, but is a holistic concept that embraces and recognises history, relationships, the material and immaterial and encompasses physical, spiritual, historical and cultural aspects (Bodkin-Andrews et al., 2016; Fredericks, 2013; Pelizzon & Kennedy, 2019). Aboriginal academic Emma Lee takes this concept a step further in acknowledging her Country as the "totality of emotive, physical, intellectual and metaphorical connections that has its own agency and influences", and by including her Country as a co-author (tebrakunna country & Lee, 2017, p. 95). For Indigenous peoples, connection to this concept of country is far more than being "on the land" or "on country". It is a complex metaphysical concept that has widespread similarities, yet at the same time is grounded in location-specific diversity. This provides a very unique opportunity as an authentic element of attraction for potential tourists.

Aotearoa New Zealand and Māori ceremony: *Pōwhiri* and *haka*

Like Australia, Māori are called upon to undertake ceremonial duties in the interests and service of cultural diplomacy in the public, private and traditional contexts. The denouement of the movie "Mt Zion" (2013, dir. Tearepa Kahi) set in the late 1970s Aotearoa New Zealand has our hero (played by music star Stan Walker) performing a traditional Māori welcome to a guest revealed, to the hero's great surprise, as international reggae star Bob Marley. Marley did tour Aotearoa New Zealand in 1979 and was formally welcomed onto a *marae*, a traditional communal space protected by each local Māori community. In addition to Bob Marley, David Bowie also received a "traditional" welcome and, on being told he would be expected to sing, quickly wrote a short song which he performed for the enthralled locals (O'Neil, 2015). The ceremony revolves around a challenge by the hosts who undertake a sequence of calls and actions that culminate in the guests being embraced and fed (Walker, 1996). Through these ceremonies, guests are taken from the status of strangers to that of friends and pseudo-family.

The imagery of these welcomes are reasonably familiar to several global audiences, particularly rugby fans, who for several generations have variously observed in silence or with cheers or jeers as the highly successful All Blacks "challenged" their opponents; Royal watchers would likewise recognise it as a very traditional process of welcome for visiting royalty. Māori cultural "performances" have been a feature of contact with non-Māori since first contact. The bloody engagements with Dutchman Abel Tasman's crew in 1642 and with Englishman James Cook in 1769 were presaged by European ignorance and Māori efforts of welcome (O'Malley, 2012; Salmond, 1991). These performances have since become *de riguer* as Māori performing groups open multiple sporting, political and cultural events and performers have been a part of diplomatic missions for many years as well as being front-and-centre for military parades. The ubiquity of welcome ceremonies has led to calls for their curtailment by racists (Curchin, 2011) and limitations by Māori.

While subject to some of the same criticism as given by their trans-Tasman neighbours, these ceremonies have also assumed a significant marker of cultural recognition. Duncan and Rewi (2018, p. 4) argued that while Māori are seen as "adventurous and highly adaptive opportunists", Māori society is also risk adverse where rituals "manage risk and mitigate any potential harm to ensure the survival and wellbeing of the community". Many rituals were designed to "ensure cultural, physical and psychological safety during times of uncertainty and conflict, and especially during intertribal engagement" (Duncan & Rewi, 2018, p. 122). While many tourists (and probably many Pākehā New Zealanders) would gloss Māori ceremonies under a mislabelled notion of "the *haka*", the *haka* itself is a specific genre of song or chant that "conveys a *kaupapa*, a message ignited by the kā, the spark and fuelled by the hā, the breath which is the actualization of mauri, the life force" (Smith, 2017, p. 12). It is perhaps the *pōwhiri*, the traditional ceremony of welcome, which is the epitome of this risk management.

The *pōwhiri* is a structured ceremonial engagement with considerable improvisation through formal speeches, songs and movements, including *haka*. The sequence of actions and words sees visitors challenged, called, spoken to, listened to, held and fed; each stage involves careful manipulation of what is permitted and what is not or not yet. Respect and reciprocity and the need to perform and articulate these values govern this ritual, as do many other Indigenous ceremonies. Despite being tradition, such ceremonies can and have been "modernised" and Māori tourism has a long history with diverse examples of such modern expressions. Tahana and Oppermann (1998) argued that cultural performances were "tailoured" towards the specific setting and needs of the clientele, with modern repertoires including "more exciting elements…of greater interest to the tourists" (p. 23). Māori, like other Indigenous Peoples, realised quickly that capitalism, while difficult to master, is conceptually simple.

One key "mega" event serves to illustrate the balance between tradition and modern cultural adaption (for want of a better term). The opening of the 2011 Rugby World Cup, hosted by the NZ Rugby Football Union, featured an extensive and complex choreography as the showpiece of the ceremony that incorporated Māori mythology and rituals. The key moment was the lone female voice chanting the *karanga* that signals the formal start of a *pōwhiri*. This then led into a digital display of what were a sequence of Māori myths on the discovery and foundation of Aotearoa New Zealand "intersected with displays of *haka* … topped off with three tonnes of fireworks" (Jackson, 2013, p. 848). While purists may cringe, the performance was recognisably "traditional" through its structured sequencing and roles manifested by (hopefully well paid) Māori performers of great skill and knowledge. It was a significant statement of territorial rights by an Indigenous community to thousands of visitors physically present and a global television audience of millions.

The national Rugby team, the All Blacks, were embroiled in contractual requirements that lead to criticisms of commodification of, among other things, the *haka* known as "*Ka mate*" (Jackson, 2013). Yet it should be noted this *haka*, most often associated with male professional athletes, has a radical reading that diminishes the overtly masculine and elevates the position of Māori women. Palmer (2016) presented this compelling history and reading of *Ka Mate*, integrating the historical context of the event, the role of women and their ability to nullify the incantations of the pursuers that saved the life of the hero. Timu (2018, p. 109) records a research participant as saying: "Non-Māori [New Zealanders] have no other way to distinguish themselves as New Zealanders other than with *haka*". On occasions, the male buttocks are (how shall we say) more than evident, and this too has its cultural logic in the *whakapohane*, which involved the deliberate baring of the gluteus maximus as an insult, such as performed to the visiting Royals (Diana and Charles) by Te Ringa Mangu Mihaka (Mihaka, 1984) in 1983.

The flourishing of domestic New Zealand tourism during COVID will undoubtedly provide yet more evidence of Māori ceremony through welcomes, hospitality and farewells. Environmental management is also seeing greater assertion of cultural approaches to contamination, accident and disaster (Ataria et al., 2018). Policy and legislative changes are also opening more political-economic space for Māori, even if this is often halting and still bitterly opposed by some. But what can be observed is the embodiment of a culture that is welcoming strangers in a way that signifies the presence of traditional (if sometimes contested) landowners and signifies a certain (if still partial) authenticity of experience.

Ceremonies and socialising tourism

In Australia and Aotearoa New Zealand, Indigenous ceremonies of "welcome" that recognise the coming together of groups of peoples have been central to maintaining, reviving, reclaiming, performing and displaying culture for many thousands of years. While they maintain specific vital elements of ancient customs, they also reflect the changing world in which they take place. The sense of connection to land is at the heart of these contemporary ceremonies. Rather than focus on perceptions of separation and "not belonging", we contend that these ceremonies act as a vehicle for embracing people, culture and history. As Emma Kowal suggests, "the welcome ritual is a vessel for issues of recognition and belonging that swarm just beneath the surface of the nation" (Kowal, 2010, p. 15).

In addition, these practices further underline the role of diplomacy in contemporary interactions between groups. In our increasingly globalised world, such diplomacy is vital in both recognising cultural difference and acknowledging historical precedents. Indigenous Peoples enact highly attuned rituals and ceremonies that reflect ancient philosophies and pragmatism when they engage with other peoples. Such practices retain their relevance in contemporary diplomacy, particularly through trade, and have increasing value as diplomatic missions struggle with sovereignty and international relations. For states with Indigenous communities, Indigenous diplomacy offers a pathway to a more harmonious world.

In the case of tourism and tourism development in the 21st century, these ceremonies can add a rich layer to the visitor experiences as well as contribute to socialising tourism by making it more "responsive and answerable to the society in which it occurs" (Higgins-Desbiolles, 2020, p. 617). By involving local Indigenous communities in welcoming tourists and guests, we can enhance the connection between not only hosts and guests, but the connection between guests and culture, place and history. Visitors can feel more connected to the place they are visiting, and get a greater knowledge, understanding and appreciation of both its historical and contemporary significance. Additionally, host groups can feel culturally empowered through authentic control and the ability to both share their culture and have a degree of sovereignty over the message being sent to tourists. This creates not only a better visitor experience, but also empowers local communities to engage with and be involved in tourism development that can contribute to their economic, social and cultural well-being.

References

Ajifowoke, M. G. (2020). Here are four positive effects of the COVID-19 pandemic, *Ventures Africa. SyndiGate Media Inc.* Retrieved 26 February 2021, from https://venturesafrica.com/covid-19-here-are-some-positive-lights-amid-the-devastating-pandemic/.

Ataria, J., Mark-Shadbolt, M., Mead, A. P., Prime, K., Doherty, J., Waiwai, J., Ashby, T., Lambert, S., & Garner, G. O. (2018). Whakamanahia Te mātauranga o te Māori: empowering Māori knowledge to support Aotearoa's aquatic biological heritage. *New Zealand Journal of Marine and Freshwater Research, 52*, 467–86.

Beier, J. M. (Ed.). (2009). *Indigenous Diplomacies.* New York: Palgrave Macmillan.

Bodkin-Andrews, G., Bodkin, A. F., Andrews, U. G., & Whittaker, A. (2016). Mudjil'Dya'Djurali Dabuwa'Wurrata (How the White Waratah Became Red): D'harawal storytelling and Welcome to Country "controversies". *AlterNative, 12*, 480–97. https://doi.org/10.20507/AlterNative.2016.12.5.4

Bolt, A. (2010). I need no welcome to my own land. Retrieved 17 March 2010, from https://www.dailytelegraph.com.au/blogs/andrew-bolt/column--i-need-no-welcome-to-my-own-land/news-story/fc4732aa3ec0c234fe64947e05137b9f

Butler, R., & Hinch, T. (2007). *Tourism and Indigenous peoples: Issues and implications*. Oxford: Butterworth-Heinemann.

Curchin, K. (2011). Pākehā women and Māori protocol: The politics of criticising other cultures. *Australian Journal of Political Science, 46*, 375–88. https://doi.org/10.1080/10361146.2011.595386

Dekker, S. W. A. (2020). Safety after neoliberalism. *Safety Science, 125*, 104630. https://doi.org/10.1016/j.ssci.2020.104630

Duncan, S., & Rewi, P. (2018). Ritual Today: Powhiri. In M. Reilly, S. Duncan, G. Leoni, L. Paterson, L. Carter, M. Ratima, & P. Rewi (Eds.), *Te Koparapara: An introduction to the Máori world* (pp. 121–124). Auckland: Auckland University Press.

Fredericks, B. (2013). 'We don't leave our identities at the city limits': Aboriginal and Torres Strait Islander people living in urban localities. *Australian Aboriginal Studies, 2013*(1), 4–16.

Gonzalez Tirados, R M. (2011). Half a century of mass tourism: Evolution and expectations. *The Service Industries Journal, 31*, 1589–601. https://doi.org/10.1080/02642069.2010.485639

Hall, C. M. (2003). *Introduction to tourism: Dimensions, and issues*. Frenchs Forest, N.S.W: Hospitality Press.

Hall, C. M. (2008). *Tourism planning: Policies, processes and relationships* (2nd ed.). London: Prentice-Hall.

Higgins-Desbiolles, F. (2006). More than an "industry": The forgotten power of tourism as a social force. *Tourism Management, 27*, 1192–208. https://doi.org/10.1016/j.tourman.2005.05.020

Higgins-Desbiolles, F. (2020). Socialising tourism for social and ecological justice after COVID-19. *Tourism Geographies*, *22*(3), 610–223. https://doi.org/10.1080/14616688.2020.1757748

Jackson, S. (2013). Rugby World Cup 2011: Sport mega-events between the global and the local. *Sport in Society, 16*, 847–52. https://doi.org/10.1080/17430437.2013.791157

Kahi, T. (2013). Mt. Zion. Small Axe Films.

Kearney, M. S. (2017). How should governments address inequality? Putting Piketty into practice. *Foreign Affairs, 96*, 133.

Kowal, E. (2010). Welcome to country? *Meanjin, 69*, 15–17.

Kowal, E. (2015). Welcome to country: Acknowledgement, belonging and white anti-racism. *Cultural Studies Review, 21*, 173–204.

Kunnie, J. E., & Goduka, N. I. (Eds.). (2006). *Indigenous Peoples' Wisdom and Power: Affirming Our Knowledge through Narratives*. Hampshire: Ashgate Publishing Limited.

Langton, M. (2018). *Welcome to country: A travel guide to Indigenous Australia*. Richmond: Hardie Grant Travel.

Lickorish, L. J., & Jenkins, C. L. (1997). *Introduction to tourism*. Jordan Hill: Taylor & Francis Group.

MacCannell, D. (1973). Staged authenticity: Arrangements of social space in tourist settings. *American Journal of Sociology, 79*(3), 589–603.

McIntosh, A. J., & Zahra, A. (2007). A cultural encounter through volunteer tourism: Towards the ideals of sustainable tourism? *Journal of Sustainable Tourism, 15*, 541–56. https://doi.org/10.2167/jost701.0

McKenna, M. (2014). Tokenism or belated recognition? Welcome to Country and the emergence of Indigenous protocol in Australia, 1991–2014. *Journal of Australian Studies, 38*, 476–89. https://doi.org/10.1080/14443058.2014.952765

Merlan, F. (2014). Recent rituals of Indigenous recognition in Australia: Welcome to country. *American Anthropologist, 116*, 296–309. https://doi.org/10.1111/aman.120

Mihaka, T. R. M. (1984). *Whakapohane: i na tuohu koe me mea hei mainga tei tei.* Porirua: Ruatara Publications.

Minnaert, L. (2016). Social tourism: From redistribution to neoliberal aspiration development. In J. Mosedale (Ed.), *Neoliberalism and the political economy of tourism* (pp. 117–128). London: Routledge.

Mosedale, J. (2016). *Neoliberalism and the political economy of tourism.* London: Routledge.

Mowforth, M., & Munt, I. (2015). *Tourism and sustainability: Development, globalisation and new tourism in the third world.* London: Routledge.

Nelson, B. (2020). The positive effects of covid-19. *BMJ, 369*, m1785. https://doi.org/10.1136/bmj.m1785

O'Malley, V. (2012). *The Meeting Place: Māori and Pakeha encounters, 1642–1840.* Auckland: Auckland University Press.

O'Neil, A. (2015). David Bowie makes Porirua marae visit. *Dominion Post*, April 15. Retrieved 24 February 2021, from https://www.stuff.co.nz/dominion-post/news/67704866/david-bowie-makes-porirua-maraevisit#:~:text=Pop%20superstar%20David%20Bowie%20made,in%20Porirua%2C%20on%20November%2023.&text=The%20trip%20was%20a%20secret,catch%20a%20glimpse%20of%20him

Palmer, F. R. (2016). Stories of Haka and women's Rugby in Aotearoa New Zealand: Weaving identities and ideologies together. *The International Journal of the History of Sport, 33*, 2169–84. https://doi.org/10.1080/09523367.2017.1330263

Pelizzon, A., & Kennedy, J. (2019). "Welcome to Country" and "Acknowledgment of Country": (Re)conciliatory protest. *Contention, 7*, 13–28. DOI: https://doi.org/10.3167/cont.2019.070103

Penberthy, N. (2016). 40 years of the 'modern' welcome to country. *Australian Geographic* (Online). Retrieved 5 August 2020, from https://www.australiangeographic.com.au/topics/history-culture/2016/03/richard-walleys-welcome-to-country/

Peters, A. (2017). *Moondani Yulenj: An examination of Aboriginal culture, identity and education.* PhD Artefact and exegesis, Swinburne University of Technology.

Ryan, C., & Aicken, M. (Eds.). (2005). *Indigenous tourism: The commodofication and management of culture.* Oxford: Elsevier.

Salmond, A. (1991). *Two worlds: First meetings between Māori and Europeans, 1642–1772.* Auckland: Viking.

Scantlebury, A. (2014). Black fellas and rainbow fellas: Convergence of cultures at the Aquarius arts and lifestyle festival, Nimbin, 1973. *M/C Journal, 17.* https://doi.org/10.5204/mcj.923

Segreto, L., Manera, C., & Pohl, M. (Eds.). (2009). *Europe at the seaside: The economic history of mass tourism in the Mediterranean.* New York: Berghahn Books.

Smith, V. (2017). Energizing everyday practices through the indigenous spirituality of haka. *Journal of Occupational Science, 24*, 9–18. https://doi.org/10.1080/14427591.2017.1280838

Stats NZ. (2019). *Māori tourism businesses employed more than 11,000 in 2019* (Online). Wellington: Statistics New Zealand. Retrieved 23 January 2020, from https://www.stats.govt.nz/news/Máori-tourism-businesses-employed-more-than-11000-in-2019

Tahana, N., & Oppermann, M. (1998). Māori cultural performances and tourism. *Tourism Recreation Research, 23*, 23–30. https://doi.org/10.1080/02508281.1998.11014816

tebrakunna country, & Lee, E. (2017).Performing colonisation: The manufacture of Black female bodies in tourism research. *Annals of Tourism Research, 66*, 95–104. https://doi.org/10.1016/j.annals.2017.06.001

Timu, N. (2018). *Ngā tapuwae o te haka – Māori perspectives on haka in sport.* Master's thesis, Otago University, New Zealand.

United Nations (2007). *United Nations declaration on the rights of Indigenous Peoples* (Online). New York: United Nations. Retrieved 13 May 2011, from http://www.un.org/esa/socdev/unpfii/en/declaration.html

UNWTO (2010). *UNWTO tourism highlights: 2010 edition* (Online). Retrieved 3 January 2020, from https://www.e-unwto.org/doi/pdf/10.18111/9789284413720

UNWTO (2015). *UNWTO tourism highlights 2015 edition* (Online). United Nations World Tourism Organisation. Retrieved 16 May 2016, from http://www.e-unwto.org/doi/pdf/10.18111/9789284416899.

UNWTO (2020). *World tourism barometer: Excerpt, 18*(1). Retrieved 5 February 2021, from https://www.unwto.org/world-tourism-barometer-n18-january-2020

Walker, R. (1996). The meeting house. In R. Walker (Ed.), *Nga Pepa a Ranginui: The walker papers* (pp. 31–51). Auckland: Penguin.

Zambrano-Monserrate, M. A., Ruano, M. A., & Sanchez-Alcalde, L. (2020). Indirect effects of COVID-19 on the environment. *Science of the Total Environment, 728*, 138813.

Zygadlo, F., McIntosh, A. J., Matunga, H. P., Fairweather, J. R., & Simmons, D. G. (2003). *Māori tourism: Concepts, characteristics and definition.* Christchurch: Tourism Recreation Research and Education Centre, Lincoln University.

2 Toxic tourism at Tar Creek

The potential for environmental justice and tribal sovereignty through Indigenous-led tourism

Bobbie Chew Bigby and Rebecca Jim

Introduction

Prayer in Quapaw language, kindly shared by Quapaw Elder Grace Goodeagle

itate wakata, kanike. xtawadade anabahowe. manika maxe ni wadaki na. daxewadadc awatawe. dedo akistowe ata iha manika akotawe desisikawe. e kikaxe owadaka, wadakitowe akodawe. washkawe tiati eko pa na. xtaadidawe, ata nikashika zani akakitowawe na, eko wadashkoze na.	Father, Thank you. We know of your love for us. You have given us earth, sky and water. We pray you would have pity on us. We are gathered to talk about our Mother Earth, and the abuses being done to her and pray for your help as we listen and learn what can be done to repair the damage. We remember those who lived long ago. They were strong. We love you and will continue to care for ourselves and all people, as you have taught us.

For the Tribal Nations that have called the state of Oklahoma home, creeks and streams are sacred places. Throughout the year, these waterways are habitats for diverse animal and plant relatives, along with serving as gathering spots for human communities and activities. Warmer months in particular bring plenty of families down to the creeks for fishing and subsistence gathering and provide the ideal spot for picnics and for children learning to swim. Above all, these living bodies of water are places of prayer, where generations of Native peoples have been taught to "go to water to pray" and

DOI: 10.4324/9781003164616-2

collect water for sweat lodges and other ceremonial activities. While the deep, verdant landscape of northeastern Oklahoma is filled with numerous waterways that crisscross this prairie geography, one creek in particular remains deeply sick, running an unmistakable rust red colour that stains the nearby rocks and land. Some locals say it looks like she is coughing and choking up blood from her wounds. This creek tells an incredibly important story of pain and resilience.

Known as Tar Creek, this toxically assaulted body of water runs through several communities and towns, including some that no longer host people due to the high levels of ecological devastation. The rusty red colour of Tar Creek is not the only marking of a sickened landscape—in fact, much of this 47-square-mile area is plagued by extensive topsoil contamination and toxic chat piles that form massive hills, along with severe subsidence risks causing many parts of the land to cave in on itself without warning. While this research tells the story of Tar Creek, at its core this work highlights the Tar Creek Toxic Tour efforts that have been organised over the years through the Local Environmental Action Demanded (LEAD Agency or LEAD) and its co-founder/executive director Rebecca Jim (Cherokee Nation). This study utilises these Tar Creek Toxic Tours as a case study to both reinforce and expand upon the claims of Pezzullo (2007) that toxic tours enable presence and connection, giving way to agency, power negotiations and empowerment. Through the lens of Tar Creek Toxic Tours, empowerment and agency are explored in more depth, particularly as these concepts apply to Native Nations, tribal sovereignty and Indigenous understandings of connection to land and water. This chapter concludes by addressing the ways in which critical engagements with toxic tourism can enable a transformation of tourism itself, from tourism as a source of environmental degradation, toxicity and structural violence (Büscher & Fletcher, 2016; Pezzullo, 2007) to one in which tourism is socialised and made relevant to the needs of the community and the environment (Higgins-Desbiolles, 2020).

The research methodology utilised in this paper is one that prioritises its engagement with Indigenous perspectives, knowledges and histories. In partnering with LEAD's co-founder, Rebecca Jim, to share the stories in this chapter and interweave them into the scholarly dialogue on toxic tourism, this research adopts a critical Indigenist methodology that first and foremost "begins with the concerns of Indigenous people" (Denzin et al., 2008, p. 2) and embeds Indigenous voices into all steps of the research process by utilising a collaborative and engaged approach.

Literature review

According to Pezzullo, toxic tourism can be understood as "noncommercial expeditions into areas that are polluted by toxins" (2007, p. 5). While so many places throughout the globe – whether inhabited by humans or not – can be considered to have identifiable levels of contaminants and pollutants, the destinations at the heart of toxic tours are generally known to be environments that have been toxically assaulted, with almost irrevocable damage to their landscapes and inhabitants. Even though toxic tours have been documented as a form of advocacy and a learning tool about environmental issues since the late 1960s (Pezzullo, 2007), toxic tours have not received a great deal of attention in the academic literature until more recently. Even less examined is the inclusion of local viewpoints from residents of toxic tourism landscapes, particularly from Indigenous and other people of colour who so often comprise the communities that have borne the brunt of the toxicity – both physically and emotionally.

Phaedra Pezzullo's book on toxic tourism provides important and unique insight into this growing phenomenon and the ways that it has developed across three different communities throughout the United States as a form of "advocacy tourism" (2007, p. 49). Pezzullo understood toxic tours as many different things and multilayered processes, including as a tactic of agency, a negotiation of power and as "embodied rhetorics of resistance aimed at mobilising public sentiment and dissent against material and symbolic toxic patterns" (2007, pp. 3–8). If agency and power negotiations are the outcomes of toxic tours, Pezzullo asserted that at the heart of what makes toxic tours such powerful processes for transformative action is their capacity to generate and communicate a sense of presence, and thus connection, for all actors alike. At one level, the toxic tours allow visitors to be present and connect with local people and also Indigenous communities who reside in these polluted communities, hearing their stories and seeing aspects of pain, discomfort and change that locals have experienced. But at another level, the tours enable visitors to experience toxically assaulted landscapes and feel a connection to these environments in tandem with the local communities that are embedded within them. These different layers of connection not only inform, but also "challenge feelings of alienation from land and each other" that are all too common in today's industrialised, urbanised and depersonalised environments (Pezzullo, 2007, p. 10). Moreover, through experiencing these environments and witnessing the interconnected impacts of toxic harm, the possibility exists that visitors have their own sense of community and ethics enlarged to include what Aldo Leopold terms a "land ethic", embracing not only humans but plants, waterways, soils, animals and all aspects of living landscapes (Pezzullo, 2007).

This term "land ethic" and the values it encompasses can be seen to have deeply rooted parallels in most Native American and Indigenous philosophies towards land and water. Winona LaDuke (White Earth Anishnaabe) (2016), a well-respected environmentalist and leader in Indian Country, wrote powerfully about traditional Indigenous perspectives on the environment and the place of human beings in relation to living landscapes, including plant and animal life forms that are considered as relations and teachers to humans. From this world view, these non-human life forms are regarded as "our older relatives—the ones who came before [us humans] and taught us how to live" (LaDuke, 2016, p. 18). In showing gratitude and acknowledgement for the lessons that these non-human relatives teach us about being in relationship with one another, humans are understood to carry the responsibility of serving as stewards, caretakers and protectors of the environments in which they inhabit. Yet in this modern age, LaDuke asserted that these relationships between humans and the living world are the very ones that industrialisation and modernity seek to dismantle, whether for purposes of profit, power or both. Following this line of understanding, LaDuke drew a direct connection between industrial activities, pollution and the subsequent loss of both bio and cultural diversity on the earth (2016, p. 22). LaDuke also serves as a leading voice in highlighting the explicit connection between settler-colonial policies towards tribal communities and the usage of tribal lands as dumping grounds for toxic waste and pollutants, whether it is on the lands of Akwesasne Mohawk, Florida Seminoles or Northern Cheyenne peoples.

While explicit studies of Indigenous-led toxic tourism are challenging to find – making this focused article a unique contribution to the discourse – the literature on Indigenous-led ecotourism provides some interesting insights that can be applied to this growing field of toxic tourism studies and the ways in which these tours intersect with Indigenous values, understandings and objectives. Moreover, perspectives on Indigenous ecotourism and Indigenous leadership within this space also underscore the ways that tourism can transform visitor understandings of community and environmental relationships while empowering Native Nations and Indigenous peoples.

As argued by both Pezzullo (2007) and Di Chiro (2000), toxic tourism can easily be understood to fall within the category of ecotourism, given toxic tourism's focus on social justice, environmental problems and the nexus between the two. Following this line of thinking, Kyle Whyte (Citizen Band Potawatomi) asserted that essential components to ensuring environmental justice is an outcome of ecotourism practiced on Indigenous lands and waters include norms of "fair compensation" and "participative justice" for Indigenous peoples and cultures involved in ecotourism initiatives (Whyte, 2010). Yet Whyte (2010) argued that these benchmarks in and of themselves are not sufficient in guaranteeing Indigenous empowerment or environmental justice objectives that ecotourism proponents so often tout as core values. He additionally advocated the inclusion of direct Indigenous participation in any ecotourism venture on Indigenous lands to harness meaningful Indigenous engagements towards an end of "environmental coalition development" (Whyte, 2010).

To illustrate the dynamics and transformative potential of this sort of direct Indigenous participation in ecotourism ventures, Freya Higgins-Desbiolles' (2009) study of Indigenous-led ecotourism and its capacity for transforming ecological consciousness is an important, instructive example. Higgins-Desbiolles utilised a case study of Camp Coorong, a Ngarrindjeri

Traditional Owner-led ecotourism enterprise in South Australia, to engage deeply with Indigenous viewpoints on ecotourism. Through Indigenous Ngarrindjeri leadership of ecotourism activities at Camp Coorong, including diverse on-country activities ranging from basket weaving with traditional materials to learning about bush tucker on hikes and telling Dreaming stories around the campfire, these ecotours were harnessed towards several simultaneous ends. The learning and dialogue that took place between non-Indigenous Australian visitors and Traditional Owners allowed for meaningful cross-cultural connections that have been recognised as part of the reconciliation process (2009, p. 149). This form of tourism engagement furthermore actively fosters a rethinking of environmental relationships for visitors, while simultaneously enabling for Traditional Owners and visitors alike "a link with the land [that] lies at the heart and soul of Ngarrindjeri culture" (Ngarrindjeri Ruwe Working Group cited in Higgins-Desbiolles, 2009, p. 61). Both Higgins-Desbiolles' (2009) and Whyte's (2010) writings recentre ecotourism as an endeavour deeply intertwined with empowerment of Indigenous communities, perspectives and practices. At their core, a common focus on relationships between humans and living ecosystems unequivocally brings together ecotourism and toxic tourism narratives together into conversation.

Background

The story of Tar Creek begins in the far northeastern corner of the state of Oklahoma, close to the nearby states of Kansas and Missouri. Historically, eastern Oklahoma had long been the traditional homelands of the Wichita and Caddo tribes, with the Osage migrating into the area later in the 1700s. Yet throughout the 1800s, what is today Oklahoma was simply land designated as "Indian Territory" by the US government. Indian Territory served as a place of relocation for tribes from throughout the country, who were rounded up and forcibly removed due to the increased expansion of Anglo-American settlers and subsequent political pressure through legislation such as the Indian Removal Act of 1830 (Clark, 2012). Of the 39 different Tribal Nations that call Oklahoma home today, 10 different tribes maintain a presence in this area of far northeastern Oklahoma. All are impacted by Tar Creek; however, it is the Quapaw Nation that features at the heart of this story.

Originally claiming their traditional homelands in what is now the state of Arkansas, the Quapaw or O-gah-pah are known as the "Downstream People" and are related culturally and linguistically to other Dhegiha Siouan tribes, including the Osage, Ponca, Omaha and Kaw (Harper, 2008). In their original encounters with the French, the Quapaw were estimated to have had a population of around 35,000. By the early 1800s, however, their numbers had decreased significantly due to displacement and diseases, and the tribe was forcibly removed to reservation land designated for them in Indian

Territory in 1834. According to Earl Hatley (Cherokee/Shawnee), the former Environmental Services Director for the Quapaw Tribe and co-founder of LEAD, their arrival into the reservation lands was marked immediately by pain and suffering (Myers et al., 2009). In a 2010 Tar Creek Toxic Tour that Hatley led, he is recorded sharing a story from their 1834 Removal of Quapaw tribal members and their entry to the Quapaw reservation jurisdiction. According to Hatley, this story was first told to him by Ed Rogers and Jeff McKibben, both former Quapaw Tribal Chairmen and highly respected Quapaw Elders who are now deceased (Figure 2.1):

> When the Quapaw reached their final area of relocation in northeastern Oklahoma, they first had to cross the Spring River. A number of tribal members died as they attempted to cross to reach their new settlement. As surviving members of the tribe stood on the shore, some claimed to see the Devil walking back and forth atop the bluff on the far bank of the river and commented that the land was cursed. The tribe agreed to leave their traditions behind if they were allowed to cross safely to their new settlement. The bluff was subsequently named Devil's Promenade, and the area of the Spring River has sometimes been referred to as the Devil's Kitchen.
>
> (Manders & Aber, 2014, p. 34)

For the Quapaw tribal members who survived the Spring River crossing, Indian Territory became their new home over the next decades. Yet due to further removals, dislocation during the Civil War, illnesses among tribal members and requirements to merge with neighbouring tribes, by 1880 only 35 Quapaws were counted on the territory (Manders & Aber, 2014, p. 35). The late 1880s witnessed the passage of the Dawes Act, entailing the forced allotment of tribal lands throughout Indian Territory by assigning parcels of land to individual tribal members as opposed to preserving communal landholdings. This period also ushered in the beginning of mining operations in the area as deposits of lead, zinc and other economically valuable

Figure 2.1 A panoramic view of the Spring River and the Devil's Promenade bluff, Oklahoma, USA. Photo by Bobbie Chew Bigby.

minerals were found, with much of the ore discovered on tribal allotments. While some Quapaw people participated in the mining activities, most did not. Sadly, even if Quapaw individuals refused to lease their lands for mining, the Bureau of Indian Affairs – the government entity charged with managing allotments and trust lands for tribal peoples – ruled individual tribal members incompetent in order for mining to proceed (Aber et al., 2010; Johnson, 2009; Myers, 2009). The legacy of allotment and subsequent mining activities for Quapaw people was short-lived wealth for a very small few, but poverty for most (Manders & Aber, 2014, pp. 35–36).

While mining operations officially started in 1891, the industry began to hum at a rapid pace by 1914 (Adcock, 2018) and quickly became an essential supplier of lead, zinc and other raw minerals for the US government's war efforts throughout the First and Second World Wars. The most productive centres of mining in the Tri-State Mining District (including Kansas and Missouri) were concentrated in the towns of Picher and nearby Cardin, both falling within what had been Quapaw tribal lands in the newly formed state of Oklahoma. Between 1885 and 1970, this region is credited for having produced nearly 460 million tonnes of ore or one-third of the raw materials used during US war campaigns (Manders & Aber, 2014). With all of the ore taken out of the ground, more than 300 miles of underground mine cavities and tunnels were left by the time the mines closed in the 1970s (Adcock, 2018). Once the ore was brought above ground, the materials were highly processed to extract the minerals, leaving behind gigantic piles of rock fragments and mine tailings, known locally as "chat piles". When the mining companies finally left town in the 1970s, no plans or efforts were made for removal or clean-up of the chat. These chat piles thus continue to dot the landscape today, serving as unmistakable evidence of the way that mining irrevocably transformed the environment and her people.

For years prior to the closure of the mines, communities throughout the Tri-State Mining District were known to have higher rates of sickness compared with neighbouring populations, particularly with tuberculosis and silicosis (Manders & Aber, 2014, pp. 29, 37, 44). Yet the true toll of these extractive industries on the land, water and people came to the forefront in the 1970s when the mining companies shut down, moved out and the pumps that removed water from the mine cavities were completely turned off. In 1979, highly contaminated underground water from the mines overflowed and spilled into Tar Creek, turning the water a distinctive, rusty orange-red colour and killing off fish and other life forms in the creek (Jim & Scott, 2007). Tar Creek feeds into the Neosho River before eventually emptying into Grand Lake, a key tourism spot in northeastern Oklahoma, and all of these bodies of water continue to be impacted to this very day. This poisoning of Tar Creek is what initially caused the area to gain the attention of the Environmental Protection Agency (EPA) and be listed by 1983 as number one on the Agency's Superfund Site National Priority List according to its Hazard Ranking Model (Manders & Aber, 2014; Myers, 2009). According to the EPA, a Superfund site is a contaminated site that is eligible for federal funding, administered through the EPA, to clean and decontaminate the site, particularly when "no viable responsible party" can be located ("What is Superfund?", n.d.). This listing on the National Priority List is a distinction that Tar Creek continues to hold into the present.

Yet the toxic presence of heavy metals was not limited to the groundwater and surface water alone. With nearly 500 million tonnes of mining waste left by the industry in the Tri-State Area, including over 83 chat piles and 63 fine tailing ponds dotting the landscape in every direction (Adcock, 2018), the true scale of the soil and air contamination was finally revealed in 1991 when blood tests of local children came back with high rates of lead in the blood. High lead exposure for any human can come with health consequences, ranging from elevated cancer rates to anaemia, but for children lead exposure can also permanently impact the nervous system and cognitive function (Karkowski et al., 2014; Neuberger et al., 2009). These profoundly terrible consequences for human health were compounded by another problem causing the area to be increasingly unliveable, namely the extent of subsidence or undermining danger. When mining activities ended, many of the pillars that had kept mining cavities from caving in were removed, resulting in an extremely unstable and regularly collapsing landscape that comprises the 47-square-mile Tar Creek Superfund Site.

While the EPA launched numerous clean-up efforts since the 1980s using the Superfund Site funding, the underlying hazards for residents of the Tar Creek area only increased with time as the funds became increasingly precarious. Subsidence risks worsened with each year and attempts to mitigate pollution through efforts such as yard remediation and diverting surface water were unsuccessful. Based on these challenges, the state of Oklahoma decided in 2005 that the only safe long-term solution was to facilitate the permanent relocation of residents of Picher and Cardin through a buyout of local homes. By 2009, nearly all of the community members had moved out of Picher and Cardin to surrounding communities. In spite of the fact that Picher and Cardin are no longer filled with residents, schoolchildren and small businesses, the former residents of these areas continue to maintain a deep sense of pride and attachment to their communities and the mining legacy of which so many families had played a critical role.

The LEAD agency and Tar Creek's toxic tours

The organisation that stands at the forefront of telling Tar Creek's story, while continuing the ongoing protracted fights for justice and remediation, is the LEAD Agency. LEAD was established in 1997 by Rebecca Jim (Cherokee Nation) and Earl Hatley (Cherokee/Shawnee) with the goals of educating and taking action on environmental degradation in Tar Creek Region and the northeastern corner of Oklahoma. For decades now, LEAD has been at the heart of supporting the impacted communities and landscape of the Tar Creek Superfund Site, with efforts ranging from delivering donations of water filters to families to providing scholarships for local students. LEAD has taken up the mantle of serving as the key advocate for Tar Creek in dealing with the government agencies at the levels of local, state, federal and Tribal Nations. In particular, LEAD has been key in continuing to keep pressure on government agencies and representatives to keep the Superfund Site programme alive and accessible. On the front of education, LEAD has hosted annual conferences for the past 22 years, while continually engaging community members and visitors alike in understanding the story of Tar Creek and other emerging Oklahoma environmental hazards, including poultry farm runoff, coal ash, asbestos and fracking, among many others.

At the centre of LEAD's efforts to educate the wider community about the Tar Creek story are the Tar Creek Toxic Tours. Rebecca Jim began these toxic tours in 1995 when she first took Earl Hatley, LEAD co-founder, on a guided drive through the 47-square-mile site to show the scale of the devastation. Hatley called this a "toxic tour". In subsequent years, countless student groups, Sierra Club members, community locals, government officials, university researchers and visitors from around the world would pile into school buses or ride bicycles that traced a fairly similar route over this damaged landscape as participants in Tar Creek Toxic Tours. To this day, these toxic tours of the Tar Creek Superfund Site are powerful and unparalleled tools in LEAD's efforts to educate and advocate for these landscapes, waterways and communities that have undergone severe trauma.

"It's about justice": a reflective analysis on toxic tours at Tar Creek over the years

> Get this. For over 25 years I have been giving Toxic Tours of the Tar Creek Superfund Site. Every element on the first tour remains on the last tour. The stops are the same. The damage remains for the most part. We are related to these degraded places that made people far from these sites wealthy.
>
> (Rebecca Jim, personal communication, 15 October 2020)

On entering the 47-square-mile Tar Creek Superfund Site on a toxic tour, visitors are often initially overwhelmed by the otherworldly features they are seeing and the scale that they cover – whether it is towering piles of "chat" mine tailings, waters that run a bubbling rust red or an eerie neon green from sewage mixing in or the simple reminders to "watch your step" and "follow the person in front" for fear that the entire ground around you will fall in due to extreme subsidence risks. But, as Rebecca Jim noted above at the start of the tour, another remarkable fact of this toxic tour is that it has largely not changed from the first toxic tours that were given back in the mid-1990s. Visitors to the area are greeted by the same, if not slightly more severe, levels of ecological devastation that visitors had encountered at the beginning of LEAD's tours, as EPA clean-up efforts across numerous Operable Units have proven ineffective overall. The act of bearing witness to the devastated landscape, often characterised as a toxic moonscape, and former communities now turned into ghost towns are a powerful way of being present with the environments and people at the centre of the Tar Creek story. It is also the first step in approaching and understanding how it is possible that visitors from near and far alike can find connection and be in relationship with "these degraded places that made people far from these sites wealthy" (Rebecca Jim, personal communication, 15 October 2020). Rebecca explained:

> We have a lot of people [in this area] with kidney disease. We have a lot of people with cardiovascular disease. We have children that are being exposed before they are even born. [From serving as a local school counselor] I have former students that are dead already. It's robbery. It was all preventable. This shouldn't have happened. And now their children will die as well if we don't finish the cleanup.
>
> (Bender, 2018)

While the ecological devastation remains largely intact from the first set of tours into the present, one of the paramount differences is the absence of people and living communities in the toured areas. Due to the voluntary buy-out programme for families living in Picher and Cardin, by 2009 most people had moved away from these towns at the centre of the Superfund Site and relocated to new, safer housing and school systems that would decrease the exposure of their children and family members to unsafe lead levels. Yet even though local people no longer roam the streets of Picher and Cardin, they still feature at the very heart of the Tar Creek Toxic Tour stories (Figure 2.2).

Having served as a high school and Indian student education counsellor in the region for decades, Rebecca Jim has seen first-hand the devastating ways that her students and countless local children were impacted by lead toxicity, ranging from high rates of attention deficit hyperactivity disorder (ADHD) and cognitive disabilities to hearing impairment, cancers and impaired decision-making leading to incarceration, teen pregnancy and other difficulties. Local people, often former Picher residents, who often accompany the toxic tours to help Rebecca and Earl share stories, are generally descendants of miners, belong to local tribes or are intermarried with tribal members or are in some way connected to the region's mining heritage and its legacy. These very people are thus the same ones whose families have been shaped by the painful legacy of lead toxicity and subsequent health impacts, but who also often maintain a deep sense of pride and connection to these former towns, along with the good memories and hard work ethic that they inspired. Standing face-to-face with local people on this tour, absorbing their stories and being exposed to their complex experiences

Figure 2.2 LEAD co-founder Earl Hatley stands on a bridge and points to the highly contaminated, rust-coloured water of Tar Creek during a toxic tour. Photo credit: Clifton Adcock.

enables a deeper connection not only with the local people themselves, but with the ways that these toxic environments have permanently transformed their lives as well as the well-being, health and desires of their future descendants. These toxic tours thus simultaneously engage deeply with the past and the future generations, while powerfully generating a connection with the people and places of the present that are still struggling along a road of healing and recovery. As Rebecca Jim explained:

> While standing with Miami High School students on that bridge [during a toxic tour], all looking down at that water [where Tar Creek mixes with the acid mine water], I asked Ryan Lowell, a sophomore at the time, what he saw. He didn't even pause. He just said he saw, "an eternal flow of evil." Well, that about summed it up for the day.
>
> (cited in Jim & Scott, 2007, p. 20)

Bearing witnesses to this toxically assaulted environment and the damage it has wrought on the lives, health and well-being of people can open up emotions that are painful, disturbing and consequential. At the same time, this experience of being present and bearing witness to this connection between a traumatised environment and people can open for visitors a greater awareness and appreciation of these landscapes, the living beings associated with them and their fundamental interconnectedness and inseparability. Going a step further, toxic landscapes can be seen to be harmed, sick, near-dying and lifeless when juxtaposed with other environments that, by comparison, are alive precisely because they support life and the living, whether speaking of people, or the fish swimming in the water or the wild onions growing along the roadside. It is in peering through this lens that visitors have the opportunity to broaden their sense of the living and what exactly constitutes community to include not only local people or Tribal Nations living in these toxic sites, but potentially also encompass the non-human creatures and elements, such as waterways, birds, aquatic life, soils and more that are just as deeply impacted by toxicity. This broadening of awareness of humans towards the living landscapes and non-human beings surrounding us as well as our complex interconnection lie at the heart of numerous Indigenous world views as a foundational value upon which all other meanings arise. This viewpoint is encapsulated by LaDuke writing that "Native American teachings describe the relations all around us—animals, fish, trees and rocks—as our brothers, sisters, uncles and grandpas...This relationship to land and water is continuously reaffirmed through prayer, deed, and our way of being—*minobimaatisiiwin*, the 'good life'" (2016, pp. 18–21).

One important way in which these Native American perspectives have slowly been introduced into non-Indigenous thinking is through the "land ethic" concept, a framework that is premised upon recognition of the interconnectedness between humans and the living landscape. This ethic or sense of value calls for humans to recognise and respect the living elements around them that can be seen to form the larger surrounding "community". As illustrated from the response of Rebecca's high school student Ryan upon witnessing the toxic mine water, this sense of a land ethic immediately comes through. Ryan does not describe what he sees in a literal manner – the foamy, rust red water, the absence of aquatic life or the pollution-stained bridge. Ryan instead describes what he sees and feels through a lens of ethics. In this sickly waterway, he perceives "evil", a word loaded with unmistakable values and a moral compass that directs not only the ways a person might feel, but how they might respond. This sentiment echo's Leopold's own original words about a land ethic, when he stated: "we can only be ethical in relation to something we can see, understand, feel, love, or otherwise have faith in" (Knight & Riedel, 2008; "The Land Ethic", n.d.).

Potential for agency, empowerment and justice

> This tour is about Justice, wanting it, a place needing Justice and a place that has waited long enough to get.
>
> (Rebecca Jim, personal communication, 15 October 2020)

While toxic tours help to lay the stepping stones for fostering presence and connection between visitors and living landscapes, these tours ultimately also hold the potential for toxically assaulted communities and landscapes to achieve agency, empowerment and possibly even justice. For advocates such as Rebecca Jim and LEAD, in fact, a yearning for justice is the key underlying intention and motivation for these toxic tours. This focus on justice is indeed at the heart of all of the work that Rebecca and LEAD undertake, given that, as she states, she is advocating for a place that wants justice, needs it, has waited far too long and has endured such profound levels of trauma. The toxic tours thus serve as powerful discursive spaces capable of fostering environmental and social reflection and change, along with negotiations of power and agency:

> You will see acid mine water flowing at 1 million gallons per day into Tar Creek, tailings piles up to 200 feet tall, and a toxic moonscape that used to be a lush tall grass prairie. But you will also see the remediation efforts of the Quapaw Nation, whose lands host this mess. The Quapaw are the first tribe in the nation to receive a primary contract for remediation of a Superfund Site.
>
> (LEAD, 16 October 2020)

The Tar Creek Superfund Site presents a highly complex story, not only because of the widespread level of contamination and subsidence or the buyouts of local residents after failed clean-up efforts. At its core, the story of Tar Creek is layered and complicated because of the different actors, communities, Tribal Nations and jurisdictions involved. This complex patchwork of communities and interests involved in the Tar Creek legacy thus imply that any discussion of agency, empowerment and justice through toxic tours must also include considerations of the important ramifications of these concepts for Native Nations and tribal communities. In other words, Tar Creek toxic tours expand upon and qualify the ways that toxic tours can enable negotiations of power, justice and agency for tribes and Indigenous peoples.

Rebecca's descriptive preview of a Tar Creek toxic tour above helps to paint a picture for visitors of the landscape they will see. But just as important, it places the Quapaw Nation and Indigenous people at the centre of understanding the Tar Creek site and the toxic tour. While the Quapaw are the primary Tribal Nation that have dealt with the Tar Creek mining legacy and currently manage the ecological disaster in the present, they are in no way presented as passive, incapable victims. Instead, the Tar Creek Toxic Tours share stories of Quapaw responses and resiliency at the level of individuals, families and the Quapaw tribal government's exercise of sovereignty, and with a specific focus on the fact that the Quapaws lead all other Native Nations in being the first to hold a contract with government agencies for managing a Superfund clean-up. LEAD, which is headed by and composed of an intertribal base of support, alongside the Quapaw Nation, thus have become the primary actors in not simply telling the Tar Creek story through leading toxic tours, but serving as the drivers of reclamation, clean-up, healing and ultimately re-establishing a connection with the land and water itself. To the surprise of many, following the buyout of local residents, the Quapaw Nation opted to stay put and bought up much of the Superfund land in an effort to take control of its remediation. Former Quapaw Tribal Chairman John Berry encapsulates his tribe's sense of agency and responsibility as stewards of the landscape and the current situation in stating: "Nobody's making us do this… it's our land, and it's our crazy" (Navasky, 2016, p. 13).

Over the years that toxic tours have been run at the Tar Creek site, the Quapaw Nation and LEAD have been joined and supported by a host of other Indigenous and intertribal efforts in assisting Tar Creek. These have included the Cherokee Volunteer Society, a local high school service-learning effort that promoted awareness on lead exposure, along with the organisation called TEAL, or Tribal Efforts Against Lead. In 2019, LEAD also had the privilege of hosting the Western Mining Action Network Conference in Quapaw, Oklahoma, where the International Indigenous Caucus passed a declaration and actionable steps addressing Indigenous rights and mining activities. These different Indigenous actors have all been at the forefront of ensuring that toxic tours are used to foster agency, justice and empowerment of Native Nations while sharing values of land custodianship with wider audiences (Figure 2.3).

Figure 2.3 Green Country Chapter members of the Sierra Club stand in front of a chat pile with Rebecca Jim (second from right) on a Tar Creek Toxic Tour in 2018. Photo credit: Clifton Adcock.

Toxic tourism transforming tourism itself

> Thousands of tourists from around the world travel Route 66 every year. Consider what our tourists [coming to Oklahoma] see. They enter the state with Quapaw as the first town on their journey but they have also entered one of our nation's largest and oldest Superfund sites [here at Tar Creek] ... Tourists have earned time off and want a 'getaway.' Those people are us, too. We want to stay close to home, fish and swim in clean water and when we get home, we want to be able to wash up with clean water and sit in the yard with our children.
>
> (Jim, 2019, p. 4)

As many from within and outside the tourism academy have noted, modern tourism has incurred many negative associations. Tourism driven largely by consumer demands and wants, endless economic growth models and overlooking the complex needs and interests of local host communities and ecosystems has been described as a source of environmental degradation, structural violence and even "toxicity" (Büscher & Fletcher, 2016; Pezzullo, 2007). However, the concept of toxic tourism, as practiced and articulated through Tar Creek Toxic Tours, fundamentally disrupts these conventions and asserts the potential of tourism to transform the phenomenon for better outcomes for people and environments.

Rebecca points out that tourists following the famed Route 66 into Oklahoma are immediately greeted by nearby chat piles and other markers that announce they have entered into a large Superfund Site. This is a radical disruption of the nostalgic, charming and upbeat images of Route 66 as both a proud representation of American progress and an attractive tourism experience. Whether visitors to this area are aware of the Tar Creek story or not, they have entered into this landscape and often interact with local communities through their purchases, tours and experiences. Those visitors who choose to join a Tar Creek Toxic Tour – estimated by Rebecca to be at least 5,000 tour participants over the past 25 years – are able to engage with local stories, communities and living landscapes in a much more profound way that forges connections. Yet even if visitors to the area do not join a toxic tour or learn about the scope of the mining devastation, Rebecca indicates that in reality, the lines between tourists and local people are already blurred by their shared interests, desires and hopes at a human level. From this perspective, the local people's desires for safe and clean water and to be with family are no different from the fundamental wants of visitors for the same things. She thus writes that "those people [tourists] are us, too" (Jim, 2019, p. 4). Through the vehicle of toxic tours, these shared human desires and interests are made more tangible and brought into specific focus. This connection that is fostered helps to transform the practice of tourism towards one not focused solely on profit, growth or external demands, but instead socialises and reorients tourism based on "the rights and interests of local communities" (Higgins-Desbiolles, 2020, p. 620). At a fundamental level, toxic tours serve as a vehicle to both highlight the interconnectedness between visitors, local communities and living environments while simultaneously fostering appreciation of and direction for these interdependencies to be enacted.

Conclusion

Visitors who join Tar Creek Toxic Tours and the LEAD members who facilitate them undoubtedly come away from these experiences with much more than simply an awareness of the Tar Creek region, its history, the tribal communities in the area and the ongoing ecological devastation. Instead, the vehicle of the toxic tour itself enables powerful awareness-building and connections on multiple levels between visitors, community members and the living waterways and landscapes upon which these tours take place. These tours help to awaken a sense of empowerment and agency for local people and landscapes not only through sharing these ongoing journeys of seeking ecological and social justice, but for also sharing the unique needs, concerns, visions and parameters of sovereign Native Nations. By fostering meaningful connections between visitors, communities and environments, these toxic tours ultimately also present a pathway full of potential for the redirection of tourism and its relationships to the people, lands and waters that make it possible in the first place.

> Nee hunga (Quapaw), Water is important
> ᎠᎹ ᎤᎱᏍᏟᏧ (Cherokee), Water is life and sacred

Acknowledgements

This paper is built upon the work of LEAD Agency and particularly its founders and leaders, Rebecca Jim (Cherokee) and Earl Hatley (Cherokee/Shawnee). Gratitude to Earl for his insights and clarifications regarding Quapaw history and his long relationship working with the Quapaw Nation. Many thanks to Quapaw Elder Grace Goodeagle for the important prayer she has shared at the beginning of this paper. And much gratitude to Quapaw Elder Ardina Moore as well as Karen Hildreth, Quapaw Museum Manager, for their assistance with the Quapaw language translation above as well as all of their efforts to ensure Quapaw language is taught to new generations.

References

Aber, J. S., Aber, S. W., Manders, G., & Nairn, R. W. (2010). Route 66: Geology and legacy of mining in the Tri-state district of Missouri, Kansas, and Oklahoma. In

K. R. Evans & J. S. Aber (Eds.), *From Precambrian rift volcanoes to the Mississippian shelf margin: Geological field excursions in the Ozark Mountains field guide* 17 (pp. 1–22). Boulder, CO: The Geologic Society of America.

Adcock, C. (2018, February 23). "Contaminated, totally": A tour of Tar Creek. *The Frontier*. Retrieved 15 September 2020, from https://www.readfrontier.org/stories/on-a-tour-of-tar-creek/

Bender, A. (2018, May 10). Rebecca Jim, Cherokee environmentalist speaks at Vanderbuilt. *Indian Country News*. Retrieved 10 October 2020, from https://www.indiancountrynews.com/index.php/columnists/albert-bender/14560-rebecca-jim-cherokee-environmentalist-speaks-at-vanderbilt

Büscher, B., & Fletcher, R. (2016). Destructive creation: capital accumulation and the structural violence of tourism. *Journal of Sustainable Tourism, 25*(5), 651–67. https://doi.org/10.1080/09669582.2016.1159214

Clark, B. (2012). *Indian tribes of Oklahoma: A guide (the civilization of the American Indian series book 261)*. Norman: University of Oklahoma Press.

Di Chiro, G. (2000). Bearing witness or taking action? Toxic tourism and environmental justice. In R. Hofrichter (Ed.), *Reclaiming the environmental debate: The politics of health in a toxic culture* (pp. 275–300). Cambridge, MA: MIT Press.

Denzin, N. K., Lincoln, Y. S., & Smith, L. T. (2008). *Handbook of critical and indigenous methodologies*. Los Angeles, CA: Sage.

Harper, B. (2008). Quapaw traditional lifeways scenario. Retrieved 10 October 2020, from http://superfund.oregonstate.edu/sites/default/files/harper_2008_quapaw_scenario_final.pdf

Higgins-Desbiolles, F. (2009). Indigenous ecotourism's role in transforming ecological consciousness. *Journal of Ecotourism, 8*(2), 144–60. https://doi.org/10.1080/14724040802696031

Higgins-Desbiolles, F. (2020). Socialising tourism for social and ecological justice after COVID-19. *Tourism Geographies, 22*, 610–23. https://doi.org/10.1080/14616688.2020.1757748

Jim, R. (2015, August 11) *Rebecca Jim's August 9 MNR editorial—name change requested. Tar Creekkeeper*. Retrieved 10 October 2020, from http://www.leadagency.org/tar-creekkeeper/rebecca-jims-august-9-mnr-editorial-name-change-requested

Jim, R. (2019, February). Tour Oklahoma tour-US- from your Tar Creekkeeper. *The LEADer*. Retrieved from http://www.leadagency.org/uploads/1/2/1/3/121336053/leader_feb_march_supplement_2019.pdf

Jim, R., & Scott, M. P. (Eds.). (2007). *Making a difference at the Tar Creek superfund site: Community efforts to reduce risk*. Vinita: LEAD Agency, Inc.

Johnson, L. G. (2009). *Tar Creek—a history of the Quapaw Indians, the world's largest lead and zinc discovery, and the Tar Creek superfund site*. Mustang: Tate Publishing.

Karkowski, M. P., Just, A. C., Bellinger, D. C., Jim, R., Hatley, E. L., Ettinger, A. S., . . . Wright, R. O. (2014). Maternal iron metabolism gene variants modify umbilical cord blood lead levels by gene-environment interaction: a birth cohort study. *Environmental health, 13*(1), 77. https://doi.org/10.1186/1476-069X-13-77

Knight, R. L., & Riedel, S. (Eds.). (2008). *Aldo Leopold and the ecological conscience*. New York: Oxford University Press.

LaDuke, W. (2016). *All our relations: Native struggles for land and life*. London: Haymarket Books.

LEAD. (2020, 16 October). WMAN/LEAD agency virtual conference, addressing mining and systemic racism: Staying connected during a pandemic. Virtual toxic tour promotion email. In authors' possession.

Manders, G. C., & Aber, J. S. (2014). Tri-State mining district legacy in northeastern Oklahoma. *Emporia State Research Studies, 49*(2), 29–51.

Myers, M. (Producer), Beer, T. (Producer), & Myers, M. (Director). (2009) *Tar Creek*. [Film] Bullfrog Films.

Navasky, B. (2016, Fall). The ballad of Rebecca Jim. In *Citizen*, 2016 GIVEWITH, A Division of CBS Ecomedia Inc.

Neuberger, J. S., Hu, S. C., Drake, K. D., & Jim, R. (2009). Potential health impacts of heavy-metal exposure at the Tar Creek superfund site, Ottawa County, Oklahoma. *Environmental Geochemistry and Health, 31*(1), 47–59. https://doi.org/10.1007/s10653-008-9154-0

Pezzullo, P. C. (2007). *Toxic tourism: Rhetorics of pollution, travel, and environmental justice*. Tuscaloosa: University of Alabama Press.

"The Land Ethic." (n.d.). The Aldo Leopold Foundation (website). Retrieved 17 September 2020, from https://www.aldoleopold.org/about/the-land-ethic/

"What is Superfund?" (n.d.). United States environmental protection agency (website). Retrieved 15 January 2021, from https://www.epa.gov/superfund/what-superfund

Whyte, K. P. (2010). An environmental justice framework for indigenous tourism. *Environmental Philosophy, 7*(2), 75–92.

3 A theory of care to socialise tourism

Sandro Carnicelli and Karla Boluk

Introduction

The socio-environmental crisis we are currently facing, as the world continues to battle the global coronavirus pandemic and the climate emergency, requires radical responses and positive transformations. The neoliberal model supported by capitalism has reduced the economy to production and distribution. This model disregards impacts on the environment, and the well-being of people thus does not respond to the needs of our world and is rather unsustainable and unjust (Mosedale, 2016). Tourism is situated in this larger context and it is evident now more than ever before that the regulation of tourism impacts is essential (e.g. Higgins-Desbiolles et al., 2019), because the laissez-faire approach to tourism management is not conducive to securing long-term sustainability (Boluk & Carnicelli, 2019; Carnicelli & Boluk, 2020). While neoliberalism has overseen a steady diminution of the governments' role in tourism governance and a concomitant expansion of private sector power, recent contemporary developments requires a redirection of the pendulum. Indeed, it is necessary for governments, among other tourism stakeholders, to mutually govern for the public good of citizens and the conservation of the planet. Signalling the irresponsibility of tourism as an industry requires our attention and awareness of human suffering, inequality, and social injustices. Such urgent concerns reveal we cannot afford to remain apathetic; we must educate tourism stakeholders for a more caring industry. What we argue here is that in order to transform our current practices, we must activate our abilities to care and enact care.

In this chapter, we endeavour to contribute to Higgins-Desbiolles' (2020) call for socialising tourism. Specifically, we reflect on how we may "place tourism in the context of the society in which it occurs [...] to harness it for the empowerment and wellbeing of local communities" (Higgins-Desbiolles, 2020, p. 618). As such, we will reframe tourism towards an activity of care towards oppressed and marginalised people and the depleted natural environment. Caring, as demonstrated and enacted by suppliers and consumers,

DOI: 10.4324/9781003164616-3

should not only be gazing, alms and donations, but this caring should be deeply moving, touching and mutually empowering to those engaged in tourism. We argue that if we engage in care, tourists may begin to engage in activism, fighting against systemic social failures that imprison the oppressed in their situation and against the egocentric consumption and depletion of natural resources. In this chapter, we will first understand the concept and nature of care, we will discuss the current state of carelessness in tourism due to neoliberal practices and move towards discussing how we can help to socialise tourism if we embed the development of caring capacity in the critical pedagogy framework we previously proposed (Boluk & Carnicelli, 2019).

Care and its concretisation: a philosophical approach

The concept of care has been discussed in multiple fields, including philosophy, ecology, nursing, medical sciences, education and feminist studies. Care has been explained by Gilligan as "an activity of relationship, of seeing and responding to need, taking care of the world by sustaining the web of connection so that no one is left alone" (1982, p. 62). Feminist scholar bell hooks incorporated care in her definition of love, arguing "when we teach with love, combining care, commitment, knowledge, responsibility, respect, and trust, we are often able to enter the classroom and go straight to the heart of the matter" (2010, p. 161). To hooks, caring and love have an important role in creating optimal and transformational learning environments. In emphasising the importance of mutuality, partnership and engagement between teachers and students, transformation becomes possible (hooks, 2010).

Leonardo Boff's (2007) assessment of care was that it may serve as a response to the detrimental human impacts of our time. Specifically, care "represents an attitude of activity, of concern, of responsibility and of an affective involvement with the other" (Boff, 2007, p. 14). In his book *Essential Care: An Ethics of Human Nature*, Boff (2017) drew part of his argument from Martin Heidegger's view of caring as a fundamental mode-of-being and an ontological aspect impossible to be disregarded. In this sense, caring is fundamental to humans and essential to their being. Such considerations are further supported by Noddings, who asserted: "In care ethics, relation is ontologically basic, and the caring relation is ethically (morally) basic. Every human life starts in relation, and it is through relations that a human individual emerges" (2012, p. 771). Noddings' (2013) work detailing the importance of care identified two parties (borrowing the hyphenated emphasis from Heidegger's existential thinking regarding "being-in-the-world") essential in thinking about caring relations, that is, the "one-caring" and the one "cared-for". Specifically, Noddings' (2013) work examined the inner dynamics of caring relationships and her analysis drew attention to the reciprocal dependence between the two parties engaged in care.

In this reflection, we depart from the point that "care" is in the essence of human beings but perverted and transformed historically. As Heidegger explained in *Being and Time* (*Sein Und Zeit*): "the whole of the constitution of existence itself is therefore not simple in its unity, but shows a structural arrangement that is expressed in the existential concept of care" (2006, p. 200). Now, the transformation of the concept of care may be seen when analysing the philology of the word. Boff (2007) presented two different views regarding the origin of "care". In one view, the word is linked to the Latin word *cura* and used in contexts of love, friendship or devotion towards someone or something. In the other, the word is derived from *cogitare-cogitatus* related to contemplation, thinking and to bringing attention too. Nonetheless, in both approaches the concept of care is relational to the importance that someone or something has to the "self".

In this relational aspect of care, the idea of conviviality and living together as expressed by Ivan Illich (1973) and Hemer et al. (2020) gains relevance. In his attempt to understand the interlinked nature of both industrial processes and ecological crises, Illich (1973) reinforced the necessity of a convivial society based on a social and sharing economy combining resources and capabilities through new forms of interaction and learning. For Illich, in order to tackle the ecological crisis that is causing environmental depletion and injustice, we should learn to *live together* with nature. This living together with the environment is evidenced in Indigenous lifeways demonstrating a profound affinity with their total environment and nature that surrounds them (Salmón, 2000; Singh, 2019). Boff's work shares similar insights drawing attention to the outcomes of encountering despair which may lead to a "profound feeling at the root of the new paradigm of living together with the Earth" (2007, p. 80). Accordingly, listening attentively to Mother Earth may lead to feelings of passion for "her" and the development of essential care.

The (re-)discovery of our capacity to care becomes the (re-)discovery of our own humanity. For Boff (2007), care becomes concrete in our actions towards ourselves, the planet, the environment, the society and those excluded. However, the industrial mindset that had guided us since the 18th century reinforces an assault on our planet and its natural resources. This assault has been aggravated by neoliberal policies and increased imperialistic and neocolonial politics with the globalisation of supply chains and production processes. This approach to production and consumption has, according to Boff (2007), reduced collective consciousness towards the planet. Moreover, organisations such as the UN that should represent the collective is still "dominated by the old paradigm of imperialist nations" who have yet to discover the earth as an object that requires care and collective policy (Boff, 2007, p. 94).

The development of the 1991 *Caring for the Earth* strategy by the International Union for Conservation of Nature (IUCN), the UN Environment Programme (UNEP) and the World Wide Fund for Nature (WWF) provided a good step forward. In this document, it is stated that "humanity must take no more from nature than nature can replenish. This in turn means adopting life-styles and development paths that respect and work within nature's limit" (IUCN, UNEP & WWF, 1991, p. 8). Furthermore, people are urged to re-examine and rethink their values and their behaviour based on the capacity of a physical environment to cope with the presence of visitors without producing detectable changes to the ecosystem, also known as carrying capacity (Buckley, 1999; Simon et al., 2004). However, in the last 30 years, research has demonstrated we continuously grow our consumption of fossil fuels and non-renewable resources. The humanitarian and ecological crisis we currently face are a result of our consuming behaviour manifested in phenomena such as human mass migration (Forman & Ramanthan, 2019), extinction of species (Moritz & Agudo, 2013) and spread of viruses previously contained to pandemic levels (Ryan et al., 2019).

Facing greater evidence of climate and humanitarian crises, the UN proposed a new sustainability strategy to its members. The UN Sustainable Development Goals (SDGs) offered a framework in 2015, building on from the failed Millennium Development Goals. The SDGs encourage countries to work towards the eradication of poverty, establish socio-economic inclusion and the preservation of environment. However, the SDGs have received due criticism particularly regarding the inconsistencies between socio-economic development and environmental sustainability goals (Spaiser et al., 2017; Swain, 2017). Tourism scholars have also critiqued the SDGs and their ability to assist in solving the world's pressing problems. Specifically, explicit in the presentation of the goals is an emphasis on capitalistic tendencies further reinforcing production and consumption (Boluk et al., 2019; Hall, 2019). This emphasis does little to support co-operation between the North and South which is essential for collectively addressing poverty and climate change (Walker et al., 2019). Boluk et al. (2019) called for critical thinking and analysis of the SDG agenda in order to challenge the status quo and unsettle "business as usual"; for instance, they highlighted how the problem of overproduction and overconsumption is cosmetically addressed through solutions branded as "responsible" approaches. Clearly, ambitions alone are inadequate in responding to the world's most pressing problems. The clear inadequacy of such agendas, despite being global in scope, presses us to think and act more radically in order to address the crises with which we are confronted.

Tourism plays a significant role in current social and environmental affairs. Abundant research has demonstrated the positive benefits tourism generates to the self, for one's mental health and the general well-being of tourists (e.g. Buckley, 2020). However, there is also ample evidence about the negative impacts of tourism to the planet, the environment and the potential use of tourism as a tool for marginalisation, segregation and exclusion (e.g. Carnicelli & Boluk, 2020). In this sense, a new approach to tourism development and definition has been proposed (Higgins-Desbiolles et al., 2019), emphasising the importance of reversing power structures and placing local communities at the centre of the activity. For Higgins-Desbiolles, "it is both tourists and tourism businesses that must be socialised into supporting the ways, needs and interests of the local societies in which they tour or offer tourism services" (2020, p. 617).

The carelessness of tourism

An emphasis on economic growth demanded by the neoliberal agenda is operative across many industries, and its impact has certainly not missed tourism (Boluk et al., 2019). The ubiquitous drive towards growth-based economics is responsible for the cycle of overconsumption and overproduction, which has resulted in irreversible impacts leading to biodiversity loss (Salleh, 2010) and the contemporary climate emergency (Ripple et al., 2020). Hall's (2019) analysis demonstrated the role leading international organisations, state agencies, destination managers, corporate actors and academic institutions play in reinforcing the drive for continual economic growth resulting in environmental and social devastation. Furthermore, Robinson et al. (2019) highlighted various issues arising from neoliberal agendas and discourses defining progress as synonymous with the growth-based economy. This work recognises the harm capitalism may cause to vulnerable populations and environments prioritising profit generation over an adoption of a caring approach.

Unfortunately, as Higgins-Desbiolles (2018) pointed out, tourism is obsessed with growth which is incongruent with sustainability goals. Notably, sustainability in tourism has not been addressed holistically (Boluk et al., 2019). Marx's (1965) perception of commodity fetishism, whereby consumers demonstrate limited understanding of the goods they consume, causing irreparable harm to environments, cultures, animals and peoples, is particularly important to reflect on within a tourism context. Commodity fetishism is recognised in the discourse of behavioural addiction, specifically "binge flying", reflecting the addiction to travelling (Cohen et al., 2011). This is further accentuated by the bucket list mentality motivated by ticking destinations off a list, representing a narrow-mindedness among tourists, rather than engaging in experiences (e.g. McKay, 2014; Taylor et al., 2018). Such concerns are taken up by Caton's (2014) reflection on tourism consumers who often struggle to address the moral burden of escaping daily routines through encounters in the spaces and places of others.

Inescapably, tourism has primarily catered to the desires of privileged tourists who demonstrate entitlement in their travel choices. The magnitude of the role tourism plays in the climate crisis was clearly spelled out by Ripple et al., who directly signalled the importance of air transportation as one of 15 indicators contributing to the climate emergency and recognising that "the excessive consumption of the wealthy [in] affluent countries" (2020, p. 1) is mainly responsible for contributing to Green House Gas Emissions. Fennell (2006) postulated that our desire to travel is tethered to who we are. Specifically, the activities or trips we select are a way we narrate ourselves to particular reference groups (Boluk & Ranjbar, 2014). Commodity fetishism supports neoliberal agendas and is clearly identified in the tourism literature. Examples include touring the Other through slum tourism that some have described as "poverty porn" (Higgins-Desbiolles, 2018) and desires to engage in volunteer tourism which form part of "neoliberal moral economies" (Mostafanezhad, 2014). Such encounters empower the tourist gaze (Hollinshead, 1992), resulting in questioning the role of the industry in promoting equity (Turner & Ash, 1975), signalling concerns regarding power (Bianchi, 2009) and echoing apprehensions about the lack of morality demonstrated by consumers and producers in tourism (Weeden & Boluk, 2014).

Another tourism example reflecting the power and impacts of neoliberal rhetoric and commodity fetishism supported by both producers and consumers is the troubling phenomenon of Last Chance Tourism (LCT). LCT describes the desire some tourists have to "see it before it disappears" and may support tourism to climate vulnerable destinations; for example, Polar bear tourism in Churchill, Manitoba, and Glacier tourism in Jasper National Park in the Canadian Rockies. LCT is the ultimate example situating visitors' lack of care for climate-threatened destinations (that is, destinations which are on the verge of extinction!) as a response to a tourism market that normalises the consumption of socio-ecological decline (Groulx et al., 2019), which may contribute to the social-cultural damage created by the tourism industry. LCT, slum tourism and some offerings of volunteer tourism shed a light on the disturbing products tourism has offered to generate profit from privileged travellers disinterested in reflecting on the consequences of their actions and instead acting with care.

Beyond the examples discussed so far, the unreflective actions and behaviours of privileged travellers may also contribute to the problem of overtourism. Overtourism may occur when destinations exceed carrying capacity limits, leading to diminishing experiences for locals and/or visitors, and may result in serious consequences for some of the world's most popular destinations (Higgins-Desbiolles, 2018). Examples include Venice, Dubrovnik, Barcelona, Majorca, Bali, Machu Picchu, the Galapagos Islands and Mount Everest. Such iconic destinations are literally bursting at the seams as their communities are inching their way to or already at the point of irritation on Doxey's Index of Irritation (Higgins-Desbiolles, 2018). Such irritations are explicitly communicated in various outward demonstrations such as spray-painted signs on wheelie bins or walls in tourist destination streetscapes communicating that tourists are not welcome in the communities they are visiting.

A realisation of the irreparable harm caused by overtourism has stimulated discussions on degrowth. Degrowth offers a counter narrative supporting "radical political and economic reorganization" mutually supporting the reduction in resource and energy usage (Kallis et al., 2018, p. 291). Büscher and Fletcher (2017) refer to degrowth as post-capitalistic politics, responding to the imperative for downsizing and responding to the out of control global patterns of consumption and production. Degrowth presents an alternative to the status quo and "business as usual" approaches serving to encourage connections with the non-human world, which could support regeneration and biocultural conservation (Cavaliere, 2017). Degrowth analysis refocuses our lens on social system-based innovations that may result in well-being (Hall & Gössling, 2013). Notions of new approaches to sustainable production and consumption from post-capitalist social movements are essential to making radical transitions for improved livelihoods (Cavaliere, 2017). Degrowth thinking encourages a reordering of priorities, challenging the traditional emphasis on profit and production. Such a focus may be the first step for a more caring tourism that benefits communities and protects natural environments.

Care-centred tourism

The critiques regarding the neoliberal forms of tourism prioritising tourism growth over environments, cultures, the poor and oppressed are important in understanding contemporary carelessness of tourism. The global coronavirus pandemic has presented an opportunity to recalibrate and to rethink the industry. Bruce Poon's instabook proposes that we, who comprise the tourism industry, must *Unlearn* (as the title reveals) and reset "what travel should and could look like" (2020, p. 2). Our intention here, as critical tourism scholars, is to contribute to reimagining tourism and resisting its destructive trajectory. Building on Poon's directive to *unlearn* the traditional way tourism has been conducted, here we move on to reflect how we may begin to engage in radical transformations centred in care.

In our previous work, we considered how we may rethink tourism in light of degrowth and community equity, drawing attention to how the transportation and travel segment often pushes responsibility in terms of progressing sustainability ends onto individual tourists (Boluk et al., 2020). Our reflections noted this choice by airlines (such as KLM) is obvious, an easy approach, and lacks critical reflection on how large-scale businesses may initiate and engage in progressing sustainability goals. Building on this analysis, we contend there are a number of ways tourism stakeholders could contribute to centring care in tourism.

Initially, stakeholders (from governments to businesses; from tourists to host communities; from tourism educators to tourism students) need to be active players and hold each other accountable in transforming the industry and moving it away from excessive consumption habits. Second, we encourage an activist mentality from travellers in supporting care-centred businesses and from host communities in pushing governments in developing care-centred tourism policies. Indeed, tourism networks should work with governments to develop systems encouraging and incentivising care-driven businesses that can pursue caring examples ripe for implementation. An example is Air New Zealand's adoption of the Maori value *tiaki*, meaning to care for people and place (Boluk et al., 2020). In 2018, Air New Zealand launched a "Tiaki Promise" channel on their in-flight entertainment, explaining how visitors may demonstrate care through their conduct (by protecting nature and showing respect for Indigenous values). Further examples of care have been noted in the area of tourism social entrepreneurship, including Boluk's (2011) work showcasing membership to the Fair Trade Tourism certification in Africa and Higgins-Desbiolles and Monga's (2020) work on GOGO Events in Australia. Congruent with these examples, we must seriously consider the role of tourism social entrepreneurs in the context of community (Aquino et al., 2018) and rural tourism development (Mottiar et al., 2018) and their capacity to support a "caring turn" (Dredge, 2017) in tourism.

We call on tourism educators to consider their role in cultivating caring tourism citizens and future industry members. Musings on the traditional nature of tourism curriculum as uncritical, primarily focused on vocational training has led to discussions on the value and importance of citizenship (e.g. Bianchi & Stephenson, 2014; Caton et al., 2014). Recently, scholars have offered a pedagogical framework supporting the cultivation of critical tourism citizens (by integrating critical topics, supporting critical dialogue, critical positionality, critical reflection and critical practice) (Boluk et al., 2019, p. 870). This framework responds to calls for the inclusion of criticality in tourism (Boluk & Carnicelli, 2019; Crossley, 2017), and may empower future tourism decision makers in developing their caring capacity. Certainly, cultivating critical tourism citizens is important; however, reflection is not enough. Reflexivity that is ongoing and iterative is needed to ensure we are: (a) involving communities, (b) listening to their needs, (c) responding to the needs articulated and (d) ensuring they are benefitting from tourism practices. It is our argument here that demonstrating, activating and centring care may be one way to respond to and resist the pressures of neoliberalism and cultivate a transformational kind of tourism. Therefore, reflexive practice and education is one way we, as tourism scholars, may contribute to addressing the inimical impacts tourism faces.

Final thoughts: caring capacity for socialising tourism in a critical pedagogy curriculum

Recently, we presented a proposal for a Critical Tourism Pedagogy (Boluk & Carnicelli, 2019; Carnicelli & Boluk, 2020). Here, it is our intention to advance this concept and introduce the idea of "caring capacity" for socialising tourism as part of the curriculum. Hall and Smyth (2016, p. 22) remind us we need "a curriculum that is engaged and full of care". Here we argue that we need a curriculum that helps learners rediscover and redevelop their caring capacity; however, in doing so, we also need to be mindful. Higgins-Desbiolles (2020) signalled the concept of "care", noting that the notion of responsible tourism worked to warn actors to "be a little more caring" and that this is not sufficient for the transformation required in the tourism industry. This "caring" signalled by Higgins-Desbiolles (2020) is passive and not an active attitude towards meaningful change. The active and transformational caring we advocate in a tourism curriculum needs to drive change and specifically "make tourism responsive and answerable to the society in which it occurs" (Higgins-Desbiolles, 2020, p. 617). The transformational care should be developed in the educational process of students and embedded in the three pillars on the emancipatory approach for the tourism curriculum that we proposed: Collaboration and Shared-Power; Experiential Praxis for Freedom; and Socio-Political Critical Reflectivity (Boluk & Carnicelli, 2019). Without developing caring capacities, it becomes impossible to have a curriculum that is focused on collaboration between all the parts of the educational process and towards meaningful solutions for complex tourism problems. Developing caring capacity is to understand "where there is domination there is no place for love" (hooks, 2003, p. 128) and as a true shared-power situation between actors as well as their recognition of their privileges is a necessary step to educate towards a socialising tourism focused on the empowerment and well-being of local communities.

In proposing Experiential Praxis for Freedom as part of the tourism curriculum, we argued:

> that considering the destructive potential of the tourism industry on economic, environmental, social and cultural levels in the curriculum will shape the learning, thoughts and actions of its future members, and should be one of the anchor points for critical agency.
>
> (Boluk & Carnicelli, 2019, p. 175)

We still believe that this is the case, but here we would like to go beyond and emphasise the importance of caring for the local communities and for their natural resources as an engine for that praxis. We need to make sure students understand the importance of local communities in tourism and develop a transformational caring capacity to fight and protect them and their rights. In building their caring capacity, students may become angry against injustices, conscious about the depletion of resources and this may propel action to fight unfairness. Lastly, the element of Socio-Political Critical Reflectivity where we argued that "this reflective approach should be included in tourism curriculum in order to offer co-learners the possibilities to review their role in the society they live, as well as their role in the society hosting them as tourists" (Boluk & Carnicelli, 2019, p. 175). The Social-Critical

Reflectivity we propose is not disconnected to transformational care. In a continuous questioning of our practices as learners, educators, tourism agents and tourists, we reflect on our capacity to care and about the levels of caring in our actions as social transformation actors.

In following the critical pedagogy of Freire (1970), hooks (2003) and Noddings (2013), we believe that education is essential for the changes required to socialise tourism (Higgins-Desbiolles, 2020). Education can transform people into critical and active citizens who will demand governmental support for local businesses over multinational corporations. Fostering critical tourism citizens (Boluk et al., 2019) will also recognise the importance and benefits of enforcing tourism corporations to pay their fair amount of taxes. Education may also inform policy that would protect the rights and employment conditions for workers and encourage alternative business models, including cooperatives, social enterprises, and not-for-profit companies. Finally, education may inform and empower local communities about the public good forms of tourism to be facilitated and the importance in keeping public assets and natural resources protected from privatisation.

Boff's (2007) philosophical analysis on care binds everything, allowing the revolution of tenderness that prioritises the social over the individual. As such, to care is to focus on the improvement of quality of life of all living organisms, not just humans; "Without care the human being would become inhuman" (Boff, 2007, p. 143). To develop the caring capacity to socialise tourism is to advocate for the improvement of the well-being of communities, the reduction of ecological degradation and the empowerment of marginalised and oppressed groups. As such, enacting an ethic of care in tourism may allow us to challenge the neoliberal rhetoric of an industry focused on generating profits at all costs. To Higgins-Desbiolles and Monga (2020), the feminist ethics of care can contribute to a tourism focused on a purpose economy and the development of social enterprises mixing business and social goals that is essential for a more sustainable and fair future. In this chapter, we discussed the importance of embedding care in tourism and developing the caring capacity of people via a critical pedagogy curriculum that will shape the future of the industry. We advocate that education in an essential route in the processes of socialising tourism as proposed by Higgins-Desbiolles (2020). Future research should further develop the concept of caring capacity in the tourism industry as a contribution to activism and radical behavioural changes towards the protection of communities, environments and the marginalised.

References

Aquino, R. S., Lück, M., & Schänzel, H. A. (2018). A conceptual framework of tourism social entrepreneurship for sustainable community development. *Journal of Hospitality and Tourism Management, 37*, 23–32. https://doi.org/10.1016/j.jhtm.2018.09.001

Bianchi, R. V. (2009). The "critical turn" in tourism studies: A radical critique. *Tourism Geographies, 11*(4), 484–504. https://doi.org/10.1080/14616680903262653

Bianchi, R. V., & Stephenson, M. L. (2014). *Tourism and citizenship: Rights, freedoms and responsibilities in the global order.* London: Routledge.

Boff, L. (2007). *Essential care: An ethics of human nature.* London: SPCK.

Boluk, K., & Carnicelli, S. (2019). Tourism for the emancipation of the oppressed: Towards a critical tourism education drawing on Freirean philosophy. *Annals of Tourism Research, 76*, 168–79. https://doi.org/10.1016/j.annals.2019.04.002

Boluk, K., Cavaliere, C., & Duffy, L. N. (2019). A pedagogical framework for the development of the critical tourism citizen. *Journal of Sustainable Tourism, 27*(7). https://doi.org/10.1080/09669582.2019.1615928

Boluk, K., Cavaliere, C., & Higgins-Desbiolles, F. (2019). A critical framework for interrogating the United Nations Sustainable Development Goals 2030 Agenda in Tourism. *Journal of Sustainable Tourism, 27*(7). https://doi.org/10.1080/09669582.2019.1619748

Boluk, K., Krolikowski, C., Higgins-Desbiolles, F., Carnicelli, S., & Wijesinghe, G. (2020). Re-thinking tourism: Degrowth and equity rights in developing community-centric tourism. In C. M. Hall, L. Lundmark, & J. Zhang (Eds.), *Degrowth and tourism: New perspectives on tourism entrepreneurship, destinations and policy* (pp. 152–69). London: Routledge.

Boluk, K., & Ranjbar, V. (2014). Exploring the ethical discourses presented by volunteer tourists. In C. Weeden & Boluk, K. (Eds.), *Managing ethical consumption in tourism* (pp. 134–52). New York: Routledge.

Boluk, K. A. (2011). Fair trade tourism South Africa: A pragmatic poverty reduction mechanism? *Tourism Planning and Development, 8*(3), 237–51. https://doi.org/10.1080/21568316.2011.591152

Buckley, R. (1999). An ecological perspective on carrying capacity. *Annals of Tourism Research, 26*(3), 705–8. https://doi.org/10.1016/S0160-7383(99)00011-0

Buckley, R. (2020). Nature tourism and mental health: Parks, happiness, and causation, *Journal of Sustainable Tourism, 28*(9), 1409–24. https://doi.org/10.1080/09669582.2020.1742725

Büscher, B., & Fletcher, R. (2017). Destructive creation: Capital accumulation and the structural violence of tourism. *Journal of Sustainable Tourism, 25*(5), 651–67. https://doi.org/10.1080/09669582.2016.1159214

Carnicelli, S., & Boluk, K. (2020). Critical tourism pedagogy: A response to oppressive practices. In S. R. Steinberg & B. Down (Eds.), *The Sage handbook of critical pedagogies* (pp. 717–28). London: Sage.

Caton, K. (2014). What does it mean to be good in tourism? In C. Weeden & K. Boluk (Eds.), *Managing ethical consumption in tourism* (pp. 19–29). New York: Routledge.

Caton, K., Schott, C., & Daniele, R. (2014). Tourism's imperative for global citizenship. *Journal of Teaching in Travel & Tourism, 14*, 123–8. https://doi.org/10.1080/15313220.2014.907955

Cavaliere, C. T. (2017). Foodscapes as alternate ways of knowing: Advancing sustainability and climate consciousness through tactile space. In S. Slocum & C. Kline (Eds.), *Linking urban and rural tourism: Strategies for sustainability* (pp. 49–64). Oxfordshire: CABI International.

Cohen, S. A., Higham, J. E. S., & Cavaliere, C. T. (2011). Binge flying behavioural addiction and climate change. *Annals of Tourism Research, 38*(3), 1070–89. https://doi.org/10.1016/j.annals.2011.01.013

Crossley, E. (2017). Criticality in tourism education. In P. Bnckendorff & A. Zehrer (Eds.), *Handbook of teaching and learning in tourism* (pp. 427–38). Northampton, MA: Edward & Elgar.

Dredge, D. (2017). Foreward. In P. Sheldon & R. Daniele (Eds.), *Social entrepreneurship and tourism: Philosophy and practice* (pp. 155–72). Cham: Springer International Publishing.

Fennell, D. (2006). *Tourism Eethics*. Clevedon, UK: Channel View Publications.

Forman, F., &. Ramanathan, V. (2019). "Unchecked climate change and mass migration." In M. M. Suárez-Orozco (Ed.), *Humanitarianism and mass migration: Confronting the world crisis* (pp. 43–59). Oakland: University of California Press

Freire, P. (1970). *Pedagogy of the oppressed*. New York: Continuum.

Gilligan, C. (1982). *In a different voice*. Cambridge, MA: Harvard University Press.

Groulx, M., Boluk, K., Lemieux, C. J., & Dawson, J. (2019). Place stewardship among last chance tourists. *Annals of Tourism, 75*, 202–12. https://doi.org/10.1016/j.annals.2019.01.008

Hall, C. M. (2019). Constructing sustainable tourism development: The 2030 agenda and the managerial ecology of sustainable tourism. *Journal of Sustainable Tourism, 27*(7), 1044–60. https://doi.org/10.1080/09669582.2018.1560456

Hall, C. M., & Gössling, S. (2013*). Sustainable culinary systems: Local foods, innovation, tourism and hospitality*. London: Routledge.

Hall, R., & Smyth, K. (2016). Dismantling the curriculum in higher education. *Open Library of Humanities, 2*(1), e11, 1–28, doi:10.16995/olh.66

Heidegger, M. (2006). *Sein und Zeit*. Tübingen: Max Niemeyer Verlag.

Hemer, O., Povrzanovi´c Frykman, M., & Ristilammi, P-M (2020). *Conviviality at the crossroads*: *The poetics and politics of everyday encounters*. Cham: Palgrave MacMillan.

Higgins-Desbiolles, F. (2018). Sustainable tourism: Sustaining tourism or something more? *Tourism Management Perspectives, 25*(1), 157–60. https://doi.org/10.1016/j.tmp.2017.11.017

Higgins-Desbiolles, F. (2020). Socialising tourism for social and ecological justice after COVID-19. *Tourism Geographies, 22*(3), 610–23. https://doi.org/10.1080/14616688.2020.1757748

Higgins-Desbiolles, F., Carnicelli, S., Krolikowski, C., Wijesinghe, G., & Boluk, K. (2019). Degrowing tourism: Rethinking tourism. *Journal of Sustainable Tourism, 27*(12), 1926–44. https://doi.org/10.1080/09669582.2019.1601732

Higgins-Desbiolles, F., & Monga, M. (2020). Transformative change through events business: A feminist ethic of care analysis of building the purpose economy. *Journal of Sustainable Tourism*. https://doi.org/10.1080/09669582.2020.1856857

Hollinshead, K. (1992). White gaze, "Red people" – shadow visions: The disidentification of "Indians" in cultural tourism. *Leisure Studies, 11*(1), 43–64. https://doi.org/10.1080/02614369100390301

IUCN, UNEP, & WWF (1991). *Caring for the earth: A strategy for sustainable living*. Retrieved 1 December 2020, from https://portals.iucn.org/library/efiles/edocs/CFE-003.pdf.

hooks, b. (2003). *Teaching community: A pedagogy of hope*. New York: Routledge.

hooks, b. (2010). *Teaching critical thinking practical wisdom*. New York: Routledge.

Illich, I. (1973). *Tools for conviviality*. New York: Perennial Library.

Kallis, G., Kostakis, V., Lange, S., Muraca, B., Paulson, S., & Schmelzer, M. (2018). Research on degrowth. *Annual Review of Environment and Resources, 43*, 291–316. https://doi.org/10.1146/annurev-environ-102017-025941.

Marx, K. (1965). *Capital: A critical analysis of capitalistic production.* Vol 3. Moscow: Progress Publishers.

McKay, T. J. M. (2014). White water adventure tourism on the Ash River, South Africa. *African Journal for Physical, Health Education, Recreation and Dance, 20*(1), 52–75.

Moritz, C., & Agudo, R. (2013). The future of species under climate change: Resilience or decline. *Science, 341*, 504–8.

Mosedale, J. (2016). Neoliberalism and the political economy of tourism: Projects, discourses and practices. In Mosedale, J. (Ed.), *Neoliberalism and the political economy of tourism* (pp. 1–20). Oxford: Routledge.

Mostafanezhad, M. (2014). Volunteer tourism and the popular humanitarian gaze. *Geoforum, 54*, 111–8. https://doi.org/10.1016/j.geoforum.2014.04.00

Mottiar, Z., Boluk, K., & Kline, C. (2018). The roles of social entrepreneurs in rural destination development. *Annals of Tourism Research, 68*, 77–88. https://doi.org/10.1016/j.annals.2017.12.001

Noddings, N. (2012). The caring relation in teaching. *Oxford Review of Education, 38*(6), 771–81. https://doi.org/10.1080/03054985.2012.745047

Noddings, N. (2013). *Caring a relational approach to ethics and moral education.* London: University of California Press, Ltd.

Poon, B. (2020). Unlearn the year the earth stood still. Retrieved 7 October 2020, from https://unlearn.travel/wp-content/uploads/2020/05/Unlearn-Bruce-Poon-Tip.pdf

Ripple, W. J., Wolf, C., Newsome, T. M., Barnard, P., & Moomaw, W. R. (2020). World scientists' warning of a climate emergency. *Bio Science, 70*(1), 8–12.

Robinson, R. N. S., Martins, A., Solnet, D., & Baum, T. (2019). Sustaining precarity: Critically examining tourism and employment. *Journal of Sustainable Tourism, 27*(7), 1008–25. https://doi.org/10.1080/09669582.2018.1538230

Ryan, S. J., Carlson, C. J., Mordecai, E. A., & Johnson, L. R. (2019). Global expansion and redistribution of Aedes-borne virus transmission risk with climate change. *PLoS Neglected Tropical Diseases, 13*(3), 1–20, e0007213. https://doi.org/10.1371/journal.pntd.0007213

Salleh, A. (2010). Climate strategy: Making the choice between ecological modernisation or living well. *Journal of Australian Political Economy, 66*, 118–43.

Salmón, E. (2000). Kincentric ecology: indigenous perceptions of the human–nature relationship. *Ecological Applications, 10*(5), 1327–32.

Simon, F. J. G., Narangajavana, Y., & Marques, D. P. (2004). Carrying capacity in the tourism industry: A case study of Hengistbury Head. *Tourism Management, 25*(2), 275–83. https://doi.org/10.1016/S0261-5177(03)00089-X

Singh, N. M (2019). Environmental justice, degrowth and post-capitalist futures. *Ecological Economics, 163*, 138–42. https://doi.org/10.1016/j.ecolecon.2019.05.014

Spaiser, V., Ranganathan, S., Bali Swain, R., & Sumpter, D (2017). The sustainable development oxymoron: Quantifying and modelling the incompatibility of sustainable development goals. *International Journal of Sustainable Development & World Ecology, 24*(6), 457–70. https://doi.org/10.1080/13504509.2016.1235624

Swain, R. B. (2017). A critical analysis of the sustainable development goals. In L. Filho & W. Cham (Eds.), *Handbook of sustainability science and research* (pp. 341–56). New York: Springer.

Taylor, M., Grimwood, B., & Boluk, K. (2018). Caring for animal welfare: Volunteer tourists and captive elephant well-being in Thailand. In B. S. R., Grimwood, H.,

Mair, K., Caton, & M. Muldoon (Eds.), *Tourism and wellness. Travel for the good of all?* (pp. 71–94). Lanham, MD: Lexington Books.
Turner, L., & Ash, J. (1975). *The golden hordes: International tourism and the pleasure periphery.* London: Constable.
Walker, J., Pekmezovic, A., & Walker, G. (2019). *Sustainable development goals harnessing business to achieve the SDGs through finance, technology, and law reform.* West Sussex: John Wiley & Sons Ltd.
Weeden, C., & Boluk, K. (Eds.). (2014). *Managing ethical consumption in tourism.* London: Routledge.

4 Local participation as tourists

Understanding the constraints to community involvement in Tanzanian tourism

Kokel Melubo and Adam Doering

Introduction

According to the World Bank (2015), tourism represents 17% of Tanzania's gross domestic product and is the country's second largest industry behind agriculture. It is also the leader in foreign exchange earnings (US$2.44 billion in 2018) and accounts for over a million jobs, comprising 11.2% of total employment (World Bank, 2015). The Travel and Tourism Competitive Index 2019 report ranked Tanzania 8th in the African region and 95th globally, indicating that it has the necessary conditions to be competitive in the international tourist market. As a result, the Ministry of Natural Resources and Tourism (MNRT) (2017) in Tanzania has positioned the tourism industry as a pillar for economic growth and development. Numerous national policies demonstrate the contribution tourism is expected to make for the future development of the Tanzanian economy. For example, the Tanzania Development Vision 2025 aims to eradicate poverty by creating employment opportunities, establishing economic diversification at the national level and planning for greater productivity in the existing tourism sector (Anderson & Saidi, 2011), while the earlier Tourism Act 2008 proposed to institutionalise public–private partnerships in order to develop appropriate tourism facilities, activities and amenities that would benefit local communities throughout the region (United Republic of Tanzania, 2008). Significantly, all of these policies emphasise the importance of full involvement and participation of local communities, which, it is argued, leads to an equitable and transparent sharing of benefits of tourism (Slocum & Backman, 2011).

Tanzania's strategic tourism policies focus on putting local communities at the centre of development. The purpose of these policies is said to offer support for the needs and requirements of local communities, mainly from rural areas, to ensure social, cultural and economic benefits from tourism are achieved at the local level (Melubo, 2020). For example, the 2019 Tourism Policy emphasises that tourism must not only be a pathway for economic development, but should also improve the well-being of the citizenry as well (MNRT, 2019). This call aligns with various international

DOI: 10.4324/9781003164616-4

programmes and academic literature arguing for greater local participation and control in decision-making (Higgins-Desbiolles, 2020), especially with respect to Indigenous communities and natural resource management (Scheyvens & Biddulph, 2018). Approaches such as inclusive tourism, social tourism, responsible tourism and community-based tourism have a shared interest in diversifying and justly distributing the benefits of tourism to the peoples and communities, who have at times been marginalised by it (Higgins-Desbiolles, 2020; Scheyvens & Biddulph, 2018). These approaches critique the tourism sector for placing too much faith in neoliberal policies of profit accumulation and private ownership that have resulted in the ongoing exploitation of rural, marginalised and Indigenous communities.

The United Nations Sustainable Development Goals endorsed by 193 states in September 2015 describes local participation as encompassing the inclusion of host communities in the control and decision-making around tourism, inclusion of marginalised peoples in the consumption of tourism, improving the quality of host-guest interaction, self-representation in tourism promotion and ensuring that tourism contributes to the overall quality of life for all (Scheyvens & Hughes, 2019; Stone & Stone, 2011). The mounting calls for local participation and more inclusive tourism derive from the fact that too often tourism focuses more on creating opportunities for international visitors and corporations than on the local communities where tourism takes place (Higgins-Desbiolles, 2020; Scheyvens & Hughes, 2019). Scheyvens and Hughes (2019) add that tourism may prove meaningful for the local people if it offers dignified work, establishes positive social ties between host-guest that overcome negative stereotypes, enables the local servicing of tourist demands and encourages locals to participate in tourism spaces, which includes access to their own lands as well as traveling these spaces in ways that align with their own world views and social practices. If the local community is not genuinely involved and in control of the use of their natural and cultural resources in tourism, long-term sustainable outcomes are less likely to be achieved (Tosun, 2006).

Research suggests local participation supports long-term sustainable development strategies, increases confidence, builds trust and ownership of projects and revitalises economies (Tosun, 2006). However, inclusive tourism and local participation comes in many different forms. For example, encouraging inclusive access to places designated as protected areas for tourism and catering to the differing recreational needs and ideologies of local communities and marginalised peoples are also an important element of local participation in tourism, but commonly receive less attention (Peters & Higgins-Desbiolles, 2012; Scheyvens & Hughes, 2019; Shultis & Heffner, 2016; Stone & Nyaupane, 2020). Local visitation to these sites is essential if the benefits of tourism are to be realised (Li, 2006). Despite its persistent emphasis in tourism planning and policy in African contexts, research suggests that genuine community involvement and control that directly provides opportunities for local people to be involved and benefit from the tourism activities taking place in their environments has been difficult to practice (Stone & Nyaupane, 2020; Stone & Stone, 2011).

In this chapter, we argue this is also the case in Tanzania's Northern Circuit where visitation and tourism to protected areas has overwhelmingly been perceived by the local community as a practice enjoyed by Western "others", not Tanzanians. According to the "Tourism Statistical Bulletin 2014–2018" report, an estimated 3.5 million international tourists visited national parks in contrast to 2.2 million domestic visitors in the same period (MNRT, 2018a). The lower level of local visitation to protected area casts doubt on the assertions of national-level strategic policies that protected areas are equally accessible and beneficial for all. Lower participation in domestic tourism to protected areas is consistent with regional trends. Botswana also observed that domestic visitors to protected areas were 40% of the total tourism visits in 2013 (Morupisi & Mokgalo, 2017). This chapter shows that despite remarkable tourism growth in Tanzania over the past few decades (except in 2020 when it witnessed an unprecedented decline in international arrivals due to COVID-19), the direct benefits and involvement in tourism for many Tanzanians has been limited. We demonstrate the limits to local participation in tourism by examining the community constraints to actively participate as tourists within the popular Northern Circuit of Tanzania.

Methods

Although this chapter deals with tourism in Tanzania as a whole, specific examples focus on the Northern Circuit. The Northern Circuit is a jumping-off point for many visitors to natural parks and safari areas in Tanzania. This countryside is biologically rich, has diverse natural resources and spectacular landscapes such as the Serengeti plains, Ngorongoro Conservation Area (NCA), Tarangire, Lake Manyara, Mount Kilimanjaro and Mount Meru. Almost 70% of overseas tourism to Tanzania visits this part of the country. This region is also popular with domestic tourists, although their participation in tourism to protected areas is low.

The discussion in this chapter draws on extensive fieldwork by the first author on the constraints and benefits of local communities' engagement in Tanzanian tourism (Melubo, 2020; Melubo & Carr, 2019; Melubo & Lovelock, 2018, 2019). We reinterpret these published sources of information to develop a broad overview concerning local community participation in tourism *as* tourists and combine this analysis with other published works addressing African tourism with respect to local participation in safari hunting, tour operation and wildlife conservation in Tanzania. These sources are analysed to examine the factors hindering local visits to protected areas. It should be noted that this study took place prior to the COVID-19 pandemic, which has led to a steep decline in international tourism – the lifeblood of the Tanzania tourism industry. This has stopped the inflow of the tourist dollar into the country (Spenceley, 2020). However, this makes the discussion in this chapter even more pertinent, as Tanzania strives to rebuild its tourism economy through the greater inclusion of domestic tourists.

Domestic tourism in Tanzania's protected areas

Tanzania has set aside large proportions of land for the conservation of wildlife, with the tourism industry playing a central role in governmental strategies for the creation of protected areas. Of the 26.6% (286,118.5 km^2) of conserved land, 57,167.50 km^2 are national parks, 114,782.97 km^2 game reserves, 58,565.02 km^2 game-controlled areas and 8,292.00 km^2 conservation areas (Spenceley 2008). While National Parks and the NCA allow only non-consumptive tourism, game reserves and game-controlled areas permit a combination of both trophy hunting and photographic tourism. Protected areas have become the unique selling point of the Tanzanian tourism experience.

In 2014, 80% of international tourists to Tanzania visited protected areas, including the Serengeti, Mountain Kilimanjaro National Park, NCA and the beaches of Zanzibar for wildlife safaris (World Bank, 2015). Available data from the MNRT Tourism Division indicates 1.5 million tourists frequented protected areas for the purpose of leisure and recreation in 2018 (MNRT, 2019). Like other African countries, most visitors to the Tanzanian protected areas come from the United Kingdom, the United States, Germany, France and Italy (MNRT, 2017). To address the serious socio-economic and conservation challenges in the region, including poverty and poaching, the Government of Tanzania has directed economic resources to attracting international tourists at the expense of the domestic market and local community needs. Promotional materials in Tanzania (Melubo, 2020; Salazar, 2010) and other African contexts (Stone & Nyaupane, 2020) are overwhelmingly targeted towards international tourists, leading to misrepresentation and exclusion of local residents in tourism and helping construct a Western gaze of tourism in the region. Tourism promotion depicts Tanzania as "unforgettable Tanzania", with pristine endless plains full of abundant wildlife and exotic luxurious beaches. Salazar (2010) argued this tourism imaginary presents a colonial, nostalgic and romantic image that forms the basis of Tanzania's "safari capital" used to attract the international market. Additionally, the use of foreign languages in the production of this imaginary – mainly English, but also French and German – speaks to the privileged position of the international over the domestic market. As we will explore, these representations are discordant with the local community's world view.

Globally, domestic visitors represent 80% of world tourist arrivals (World Travel and Tourism Council, 2018). However, in Tanzania, the number of domestic visitors to protected areas is significantly lower. In 2019, 790,723 domestic tourists visited protected areas in Tanzania (MNRT, 2019). This is the case despite the important contributions domestic tourism can make to a country. In culturally diverse regions like Tanzania, domestic tourism has the potential to enhance cross-cultural communication, break down social barriers and foster national integration as citizens travel to explore their own country and interact with their fellow citizens (Kenya Government, 2020).

However, the spending power of domestic tourists is often weak, leading many to purchase inexpensive, locally owned products and services. Although lower in spending, this flow of capital goes more directly towards local small businesses and economies (Shah & Gupta, 2000; Sindiga, 1996). As a business strategy, domestic tourism appears to be more predictable, reliable and sustainable in the long term because it is less prone to external factors such as seasonality, accessibility of air transportation and the international political climate which can abruptly and radically redirect the flow of international tourist arrivals (Scheyvens, 2007). In turbulent times, like during the outbreak of the COVID-19 pandemic, domestic tourism helps a destination withstand shocks, disperses visitors geographically across regions and to less visited rural areas, sustains jobs and keeps the industry alive (Altuntas & Gok, 2020; Grančay, 2020). Evidence suggests domestic expenditures are also critical for pandemic recovery efforts of conservation areas in the African context (Kenya Government, 2020; Lindsey et al., 2020). Furthermore, domestic tourists know the place beyond the stereotypes and have greater resilience for a diverse range of crisis events, including terrorism and conflict, disease outbreaks (e.g. SARS, MERS, Ebola and COVID-19) and natural disasters (Avraham & Kitter, 2013). For example, following the post-election 2007 riots, the Kenyan government shifted marketing efforts to the domestic Kenyan market to sustain the tourism industry until foreign tourism regained confidence. Environmentally, domestic tourists are considered low carbon, in the sense they are less likely to use long-haul flights and use land-based transportation as well as local goods and services (Becken et al., 2003). Socially, domestic tourists are potentially less disruptive as the host-guest cultural distance is reduced and the social impacts are less prevalent because of shared cultural norms (Nyaupane et al., 2020).

Cognisant of the significant contribution of domestic tourism, the Tanzanian Government has developed strategies that encourage and facilitate the local population to visit their country's protected area attractions. Some of the initiatives include offering reasonably affordable packages to visit the national parks during local events and holidays such as the Dar es Salaam International Trade Fair, Farmers Trade Fair, Christmas and Easter. Packages may subsidise part of the costs of transportation, tour guiding service, refreshments and accommodation as well as free entry to parks. Additionally, Tanzania National Parks (TANAPA) has built affordable *bandas* (thatched houses), huts, rest houses and hostels in the protected areas to cater for domestic tourists (Melubo & Carr, 2019). They charge lower entrance fees for attractions compared to their foreign counterparts. For example, while entrance fees for East African adult citizens are US$5 (TSh 10,000), for the non-resident adults it is US$60 (TSh 120,000) for entry to Kilimanjaro, Gombe, Serengeti, Manyara and Tarangire National Parks (TANAPA, 2018). National campaigns such as "Tembea Tanzania, Talii Tanzania" ("Tour Tanzania, Enjoy Tanzania") have also been launched to encourage locals to explore their own country, with a special emphasis on national parks in the Tarangire, Serengeti and Mikumi. These promotional campaigns aim to grow the domestic market with the slogans: *Twende Tukatalii* (Let's go to experience), *Urithi wetu, Fahari yetu* (Our heritage, our pride) and *Utalii Uanze na Mtanzania Mwenyewe* (Tourism begins with citizens). To ensure local access, the information dedicated to domestic tourism and travel from the Tanzania Tourist Board (TTB) is also written in the local Swahili language. Furthermore, the Tanzania Broadcasting Corporation (TBC) launched the Tanzania Safari Channel, a Swahili programme, to air tourism-related programmes to boost tourism numbers and encourage locals to tour the richness of their own country.

Conservation-based tourism agencies have also organised familiarisation and educational trips for local and international journalists. Some have even become part of a communication network that updates recent happenings in Tanzanian destinations. The media networks inform the public about key attractions, special features, historical information about the places and communities, investment opportunities and contact information. Famous personalities, local celebrities and musical performers, such as the Tanzanian Bongo singers Ali Kiba, and beauty pageant competitors, have served as goodwill ambassadors to help promote domestic tourism within Tanzania. However, despite all these efforts, local participation remains limited due to a variety of reasons.

Local participation *as* tourists in Tanzanian tourism: involvement, constraints and challenges

Protected areas and tourism are foreign concepts

One significant reason for the limited engagement of locals in tourism to protected areas is that local communities in Tanzania's Northern Circuit also understand "protected areas" and "tourism" as foreign concepts (Melubo, 2020; Melubo & Lovelock, 2019). This builds on the argument that ecotourism, tourism in protected areas and even tourism itself are Western constructs (Akama, 1996; Cater, 2006; Shultis & Heffner, 2016). This makes the very idea of local involvement as *tourists* discordant with community understandings and world views of their homes, lands and social practices of travel. For instance, a recent study by Melubo (2020) observed that a significant number of local people associate tourism in protected areas as an activity undertaken by others, namely white, Western tourists. Tourism was perceived as a cultural practice of Americans, Europeans and other wealthy nations, and also that the practice of tourism requires foreign currencies.

For them, tourism is defined by its currency: the Western dollars and Euros. The reverse gaze of the local participants in the study described the home of international tourists as crowded with people, concreted with modern infrastructure and having limited wildlife. For these reasons, locals were reluctant to engage with tourism out of hesitation of being labelled a tourist.

For local communities located adjacent/within protected areas, these spaces are the home of wildlife, including big game species, and flora, but also places built for tourists. Tourism-driven protected areas comprise mostly upscale lodges/campsites located in pristine natural areas with little to no human settlement and agriculture. Previous interviews with locals in this area have repeatedly shown the establishment, upgrading, expansion and management of Tanzanian protected areas commonly excludes local residents (Melubo, 2020; Melubo & Lovelock, 2018, 2019). The "inclusion" of community land into protected areas has imposed restrictions that limit their access to ancestral land necessary to maintain their livelihoods. Hunting wildlife for subsistence and the use of these spaces for pastoral grazing, for example, was severely limited or banned, and individuals who relied on such activities were required to look elsewhere for food security (Honey, 1999).

Ongoing misrepresentation and exclusion of local residents in protected area tourism developments have established a dominant Western world view and deprived communities of their valuable resources, consequently creating distrust and fear towards tourism amongst communities in the region (Akama, 1996; Melubo, 2020; Stone & Nyaupane, 2020). As Akama (1996) noted, protected areas do not evoke images of an exotic environment for adventures, but rather conjures images of a harsh colonial legacy which kept them out of places of ancestral origin. Colonialists maintained that Africans were wiping out the wildlife population and therefore needed someone else – the West – to oversee the management of these resources and maintain "pristine" and "untouched" wilderness conditions. In Western eyes, the African wild and untamed nature is innocent and will only survive if people are kept out. Failure to do so, it is argued, will likely result in the disappearance of such pristine spaces (Neumann, 1998). The Western-centric world view of protected areas violently removed the Maasai from the Serengeti, with one claim arguing that the water reserves and land were too fragile to support their livelihoods and that for wildlife to survive a safe environment devoid of people was necessary (Honey, 1999). This historical lineage is itself a significant challenge for local participation in tourism as it serves as a considerable ideological "barrier" for local participation. Participating in tourism as a tourist would mean embodying this colonial heritage.

Socialised for socialising: joy is not found in the wild but in visiting friends

For the local Tanzanian community, the meaning of leisure is not found in nature. Visiting protected areas is seen to limit their ability to interact and socialise, which is the primary leisure practice for many Tanzanians (Melubo, 2020). Comparable to Indigenous leisure practices in Canada (Shultis & Heffner, 2016) and Australia (Peters & Higgins-Desbiolles, 2012), many Tanzanians in rural areas are not socialised to seek individualised tranquillity by reconnecting with themselves and nature in the wild. Rather, leisure time centres on socialising with family, friends and other relationship-oriented pursuits and an occasional excursion to urban centres by rural residents for shopping. This view was captured in the following statement:

> Isolating oneself is not an African thing; you don't run away from others... people like being together, meeting and talking to each other and not looking for thrill and excitement in the wild or exotic. They would rather go to malls or places of meeting.
>
> (Melubo, 2020, p. 10)

Akama (1996) and Stone and Nyaupane (2018, 2020) noted that people influenced by African values prefer to go to places to meet others for social gatherings rather than protected areas. To consider family members' opinions when thinking about travel and leisure decisions is given priority.

This demonstrates that the understanding and interests of domestic visitors differs from the Western framing. For instance, while Western tourists would emphasise privacy, tranquillity and quietness, Melubo's (2020) study indicates that most Tanzanians rarely visit protected areas for recreational purposes. Tanzanians are more comfortable in places where they can socialise and bond with others for *nyama choma* (barbecued meat) and freely drink local beer. In fact, some scholars have indicated that one of the shortcomings of tourism is the failure to grasp the concept of tourism and preferences of local tourist participation from their own geographical or cultural frames (Stone & Nyaupane, 2018, 2020). This narrow understanding of tourism reinforces a Western world view of tourism, leisure and "conservation" (Shultis & Heffner, 2016; Stone & Nyaupane, 2018). Therefore, understanding local and indigenous world views, social practices and ways of subtly rejecting tourism in protected areas is essential for building a more inclusive, just and diversified tourism industry (Peters & Higgins-Desbiolles, 2012; Stone & Nyaupane, 2018).

Protected areas are for those with dollars

Since Tanzanian tourism products are targeted at high paying international tourists, prices for tourism services are often prohibitive for local communities. Although the entrance fees to protected areas were deemed low, a lack of affordable accommodation for domestic tourists and a lack of private transport were other constraints to visiting protected areas for many citizens. In terms of transport costs, a visit to Serengeti National Park from Arusha town would include a TSh 10,000 (US$5) entrance fee, TSh 20,000 (US$10) for a motor vehicle with tare weight up 2,000 kg and TSh 400,000 for 200 litres of fuel. Most of the protected areas do not have budget accommodation (hostels and campsites) or restaurants and catering services to the needs of the local visitors. To navigate some of the parks such as the Ngorongoro Crater and Serengeti National Park, one needs a four-wheel drive vehicle. To spend such money on being a tourist was described as unthinkable, as one participant commented in Melubo's study:

> You can't think of luxury stuff [visiting protected areas] if you don't have food. Get food, school fees for kids, support *ndugu* [relatives] then think of travelling to places. It will sound like of waste and not sensible if you start spending for travelling.
>
> (2020, p. 257)

Such costs discourage locals from visiting protected areas based on their lower economic status. From the local perspective, travelling for recreational purposes is a luxury that is not within the reach of most Tanzanians. There is also a concern that much of the information about protected areas focuses on foreigners and not locals. To address this low demand, lack of information and to encourage local visitation to these areas, tourism agencies have begun educating school children about the importance and beauty of wildlife and conservation areas.

Difficult neighbours: the troubles of living with tourist protected areas

Living adjacent to or inside a tourist protected area brings both advantages and disadvantages. However, for the Indigenous Maasai community, the troubles of living with tourism protected areas outweigh the benefits (Melubo & Lovelock, 2019; Nelson, 2003). To pave the way for the protection of wildlife and international tourism, the Maasai people have lost access to prime grazing areas in Tarangire in 1970, Ngorongoro crater in 1974, Mkomazi in 1988/89 and the Arusha and Serengeti areas in 1959, all justified by the argument that they were not managing the environment correctly: accused of overgrazing and slash-and-burn agriculture. This land traditionally comprised sacred areas and water sources for residents and their livestock. Their removal equated to a loss of livelihoods, history, memory and representation. Colonial and postcolonial governments have been advocating for the relocation of people from such areas for decades, indicating that conservation and the tourism trade have enjoyed privileged positions in Tanzanian politics. Brockington (2002) suggested that Indigenous communities have experienced the brunt of modern conservation misfortunes, as between 50%

to 100% of present protected areas were once occupied by them. As Melubo and Lovelock (2019) detailed, people living inside or near a protected area are unable to gather firewood or poles, unable to walk to neighbouring villages and once resettled are not be able to live as they once did.

For example, in Ngorongoro, the 2018 Multiple Land Use Model Review Taskforce proposal would relocate 70% of the nearly 100,000 residents in the area for wildlife habitat and tourism development, with the remaining 20% to be relocated within the area but away from the prime tourist attractions (MNRT, 2019). The Model Review Taskforce is the result of pressure from international conservation bodies – including the UN Educational, Scientific and Cultural Organisation World Heritage Centre and the International Union for the Conservation of Nature (IUCN) – which called on the Tanzanian Government to find more appropriate methods of zoning and resettling local communities in the area; if put into effect, 25 administrative villages will no longer exist. The MNRT (2019) report further explained that the annexation of 3,983 km^3 of community-owned land into the conservation core zone, in Loliondo, the Natron Game Controlled Areas and Selela forests, could expand the NCA from 8,100 km^2 to 12,083 km^2. If the government approves the annexation, residents in those places would lose their rights to livelihood, including subsistence farming, access to forest products, grazing land and water use as well as suffering livestock depredation and threats to human safety (Melubo & Lovelock, 2019). Other disproportionate costs borne by people dwelling on the edge of a protected area involve not only a threat to livelihoods but also their lives (Nelson, 2003). Changed pastoral livelihood, confrontations with conservation authorities and transformed land tenure approaches from "people with wildlife" to "wildlife without people" has created animosity, altered relationships and social networks and transformed cultural values and lifestyles (Goldman, 2011; McCabe et al., 1992). Restrictions on water usage and access to pasture are a matter of basic life necessities (MacKenzie et al., 2017). Suggestions to relocate marginalised communities from ancestral lands against their will is doubly unfair and against the principles of sustainable development that encourage consent and full participation of all parties in decision-making (Higgins-Desbiolles, 2020; Saarinen, 2009).

Based on the above, rather than strengthening co-operation between protected areas and adjacent communities or truly serving the needs of the local communities, tourism in protected areas has facilitated eviction and forced the local community into increasing poverty (Monshausen et al., 2016). The rhetoric of local participation and community involvement is a discursive tool that we argue is a form of exclusion through inclusion. The Government of Tanzania encourages, facilitates and structurally supports conservation *without* local participation. Under such duress, the very thought of local participation in tourism *as* tourists is extremely difficult.

Community involvement in the hunting tourism industry: an exemplary exclusionary practice

The distinction between the industry structures and ideologies privileging international tourism practices over local world views and involvement is perhaps nowhere more clearly exemplified than with hunting tourism. Trophy hunting is an important tourism sector in Tanzania that is said to generate substantial tourism revenue and create incentives for wildlife conservation (MNRT, 2018b). Between 2009 and 2017, a total of US$163,465,077 was collected from hunting tourism (MNRT, 2018b). Tourist hunting, also known as consumptive use of wildlife, in Tanzania takes place in Game Reserves, Game Controlled Areas, Wildlife Management Areas (WMAs) and Open Areas, all under the management of the Tanzanian Wildlife Management Authority (TAWA). Problematically, the sector has been accused of undermining the participation of ordinary citizens, which directly contradicts the Wildlife Conservation Act No. 5 of 2009 and Tourism Act No. 29 of 2008 that encourage citizen involvement and provide equal rights to legal access to wildlife resources (Kiwango et al., 2015). Nearly 90% of the trophy hunting outfitters are either fully or partially foreign owned (MNRT, 2018b). In most cases, these companies are locally based, but controlled by foreign hunting interests. In terms of the tourist market, 70% of the hunters come from Western Europe and the United States, creating economic leakage in the industry. This trend is even more problematic when considering that hunting wildlife for subsistence for local communities was severely limited or banned in order to establish these trophy hunting protected areas.

To address this imbalance, the 2009 Wildlife Conservation Act section 39(3) introduced a provision that states the percentage of foreign-owned company areas allocated hunting blocks shall not exceed 15% at any one time. Despite such deliberate efforts aimed at addressing this unequal access to resources, a significant number of local communities have not been able to respond due to the large capital investment required, the knowledge (or even desire to) to be a competitive business in the international market and a poor understanding of the overall industry (MNRT, 2019). To qualify for a hunting block, applicants must have a suitable office premises for carrying on the hunting business; have a fleet of no less than two four-wheel drive pickup trucks in good running condition, registered under the company's name and inspected by relevant authority; proof of reliable communication facilities; at least six tents, two refrigerators and two freezers; and beds and other necessary furniture (Tanzania Wildlife Management Authority [TAWA], 2018). In addition, in lieu of the above equipment, a fully Tanzanian-owned hunting company may produce a Bank Bond or guarantee to the tune of at least US$300,000 as a commitment for the purchase of the equipment within three months after allocation of the hunting block (MNRT, 2018b).

In 2018, the Government introduced an auction system for allocating hunting areas (TAWA, 2018). The purpose of the system was to optimise the economic and social benefits from hunting while enhancing transparency and accountability in the hunting block allocation process. The adoption has, however, been criticised for not directly addressing the constraints to local community involvement in hunting operations, with the underlying profit-driven incentive structure once again guaranteeing wealthy and elite foreign ownership of the blocks. Local participation within the hunting tourism industry, and also *as* hunters themselves, remains highly unlikely. Despite the neoliberal, "trickle-down" argument provided by the government that auctioning hunting blocks will guarantee the highest possible revenues, thereby offer the most financial support to the largest number of Tanzanians, the reality is that only a few elite Tanzanians are able to own hunting companies. However, they are not the focus of poverty alleviation programmes prioritised in the Tanzanian government's tourism policies. In fact, the auction system would give the wealthy outfitters greater access to wildlife while further restricting local access to the same resources, once again denying community's customary usage rights to hunt for subsistence purposes either for consumption or to provide families with food, clothing, medicinal or materials for other cultural products.

"Game fees", which refers to the price paid for an animal hunted or wounded by a tourist hunter, are also prohibitively expensive for the local community, making the barriers to participating difficult to overcome. For example, the game fee for a tourist hunting a Buffalo African Cape (*Syncerus caffer caffer*) is US$2,500 and a hippopotamus (*Hippopotamus amphibious*) is US$1,500. While the amount may sound relatively affordable to some, it is excessively expensive for the citizenry to participate. Local communities may not have the financial resources to support such fees. To become a professional hunter, the regulations require Tanzanians to obtain a professional hunting license, which means paying US$1,200. Besides the prohibitive costs, one has to serve as an apprentice for two years. The outfitters, most of which are foreign-owned with headquarters in Western countries (i.e. the United States, Germany and Italy), have been reluctant to hire locals on the grounds of limited knowledge of international marketing, poor networks with foreign markets and limited field experience. The professional hunter has the responsibility of marketing hunting safaris overseas and bringing in clients for the companies. Some claim the Indigenous local candidates lack the English language skills to communicate effectively with clients (MNRT, 2018c). Furthermore, a lack of a formalised training system, certification, grading, licensing and registration of local professional hunters throughout the country has also restricted community involvement in the industry.

To be clear, the local citizenry do not necessarily want to be "hunting tourists"; they would ultimately prefer to reclaim ancestral/communal land in order to continue hunting for subsistence and to maintain cultural practices associated with the region and its wildlife. For now, however, the hunting tourism industry exists, and although involvement in the industry is not ideal, it could still afford an opportunity to maintain traditional practices and share cultural knowledge that is deeply connected with these lands. Being excluded from the hunting tourism industry is therefore a double exclusion: first through physical removal of ancestral lands and then again by being deprived of the opportunity to share knowledge both by being involved with their lands and being with others on their lands, even if through tourism opportunities. Hunting tourism is therefore exemplary of how uneven and unequal social relations are being reproduced and maintained at the expense of local community's subsistence and recreational needs and values.

Limits to community participation in tourism

Although the wildlife conservation and protected areas tourism policies of Tanzania call for the participation of local communities in the protection, utilisation and benefits extracted from wildlife resources, actual participation in tourism remains limited. A case in point is the establishment of the WMAs in 2003, which defined a form of community-based conservation that empowers villagers or communities to contribute to sustainable wildlife conservation (Tanzania Wildlife Management Authority [TAWA], 2020).

The WMA has been presented as a departure from the centralised and exclusionary approach in which the community was excluded by the use of force from the management and use of natural resources. WMA promised to bring decision-making closer to the people, thus increasing participation in tourism and bringing greater accountability to the industry. However, the establishment of WMA has been marred with a myriad of issues. Critics see the WMA as an expansion of the protected areas territories and resources, involving land use restrictions and further encroaching on Indigenous community lands. MNRT (2017) reported that the 38 functioning WMAs have now incorporated 7% of Tanzanian territory, all from the village lands. For example, many Maasai understand the WMA to be the latest invasion on their sovereignty, which has seen their land appropriated for conservation and tourism investment purposes (Snyder & Sulle, 2011). Once again, the "new" WMA focuses on conservation and less on local participation as tourists, as local landowners and as the caretakers of their ancestral land.

The establishment of WMA has been characterised as lengthy, arduous and bureaucratic, demanding technical knowledge and an understanding of laws and policies governing the wildlife resources, making it difficult for ordinary citizens to participate in the industry without the assistance of foreign experts (Kiwango et al., 2015). For example, of the 38 registered WMAs, only 13 are fully operational. Kiwango et al. (2015) reported that without the assistance of outside professional experts and financial resources from non-government organisations – such as the African Wildlife Foundation (AWF), Frankfurt Zoological Society (FZS), Honeyguide Foundation

(HGF) and Wildlife Conservation Society – (WCS) – no village could have established and/or operated a WMA. Other devastation borne by WMA villages include loss of livestock, crops, dependency on international agencies, continued state control over revenue collection, superficial community participation, conflict over land use and unequal benefit-sharing mechanisms. In this context of significant structural inequalities, "local participation" comes to mean assimilation into a managerial discourse of protected area management, with little consideration given to the recreational, subsistence or land stewardship needs of local communities. Suffice to say, deeply integrated community involvement in tourism protected areas seems to be an illusion. Decision-making power, management and ownership of natural resources lie with the government-controlled agencies such as TANAPA and TAWA. Local communities have only been given limited usufruct rights where they have the right to operate businesses, if one can afford it, but not own the land (Kiwango et al., 2015). As overall custodians of the natural resources in WMAs, the TAWA allocates hunting blocks, wildlife quotas to all hunting outfitters and unilaterally decides on the number of species to allocate to which outfitters. So, although WMAs place a lot of emphasis on the policy on giving local communities a voice in natural resources management, as Bramwell and Lane (2000) caution, such bottom-up approaches may simply be "window dressing" to avoid addressing more complex community concerns, conflicting world views and involvement in tourism.

Conclusion: socialising tourism for local tourists

Although the tourism industry has a strong foundation in Tanzania, local people and communities are yet to fully reap the benefits associated with it. Yet, to offset the fall in international arrivals, most destinations have turned to domestic tourism in the wake of COVID-19. In Tanzania, this shift to local participation in domestic tourism has proven difficult (Melubo & Carr, 2019). This chapter examined some of the constraints in Tanzania to local community participation in tourism as tourists as the tourism sector shifts its focus to the domestic market, at least as a temporary pandemic recovery measure. The fact that most local people consider the concepts of "tourism" and "protected areas" to be Western constructs, suggests the idea of local participation as tourists, in its current form and understandings, is discordant with local culture and social practices. Instead of spending their free time wild game hunting in untouched and untamed nature, many local people seek recreation in visiting and socialising with friends and relatives. This suggests that the well- promoted segment of protected area tourism does not appeal to prospective domestic travellers. Stone and Nyaupane (2018) observed that in most African and Asian societies, having fun moments with friends and family is an integral part of social and cultural life.

African protected areas are a legacy of colonial governments (Stone & Nyaupane 2018). They were created as spaces for the colonial masters to hunt and recreate. African governments such as Tanzania promoted the Western ideas of protected areas for recreation, but in doing so overlooked the importance of encouraging the local residents to visit and, in many cases, return to lands appropriated for protected areas (Neumann, 1998). The low visitation by citizens testifies to the historical setting of protected areas that served the interests of the wealthy, predominantly white, international visitors (Fletcher, 2014). The transport costs of requiring a four-wheel drive vehicle to navigate the land in a short time frame, combined with the costs for meals and accommodation in protected areas ensure that it is predominantly international tourists who benefit from Tanzania's protected areas; they are prohibitively expensive for local people to afford. The absence of local citizenry and their lack of appreciation could be major impediments to the long-term sustainability of both tourism and protected areas in Tanzania's Northern Circuit in such circumstances.

Based on the above, the following recommendations are drawn. First, wildlife activities are not favoured by many local people. There is a need to develop culturally embedded products and activities that give locals a space to socialise while recreating. Such products may include sports tourism, home stay, volunteer tourism, social tourism, agro-tourism, visiting friends and relatives (VFR) and even post-COVID recovery through public tourism (see Yamashita, chapter 10). Other ways to encourage citizen-centred tourism in protected areas is simple: provide more affordable and accessible tourism experiences for Tanzanians, such as reducing the entrance fees, providing tourist facilities suitable for Tanzanian recreational cultures and practices and putting in place campaigns to encourage Tanzanians to explore their own country as public citizens (Okello et al., 2012). At a strategic level, this means revising tourism policies, plans and visions for Tanzanian protected areas to be more inclusive of the local tourists' social practices, providing services and infrastructure better aligned with the recreational needs of the community. Socialising tourism must be inclusive; in Tanzania, this means making protected areas, quite literally, more social by focusing on collective and communal activities instead of the individualised international tourism spaces.

Rethinking the structure and ideologies of Tanzanian tourism in protected areas, to place greater emphasis on local community recreational needs, social practices and world views, is a seemingly obvious yet frequently overlooked means to increase community involvement and autonomy in what is still considered by many in Tanzania as a hegemonic Western tourism industry. In doing so, one can begin to encourage more inclusive tourism practices by ensuring local communities are also "active touring and travelling agents" – leading the way for their international counterparts (Peters & Higgins-Desbiolles, 2012, p. 76). Socialising tourism is a call for inclusive tourism. And by supporting greater local participation in tourism *as* tourists – and not only hosts or stakeholders of industry – we can begin to reimagine and reconstruct the tourism industry in a more culturally embedded and inclusive way.

References

Akama, J. S. (1996). Western environmental values and nature-based tourism in Kenya. *Tourism Management, 17*(8), 567–74. https://doi.org/10.1016/S0261-5177(96)00077-5

Altuntas, F., & Gok, M. S. (2020). The effect of COVID-19 pandemic on domestic tourism: A DEMATEL method analysis on quarantine decisions. *International Journal of Hospitality Management, 92*, 102719. https://doi.org/10.1016/j.ijhm.2020.102719

Anderson, W., & Saidi, S. A. (2011). Internationalization and poverty alleviation: Practical evidence from Amani Butterfly Project in Tanzania. *Journal of Poverty Alleviation and International Development, 2*(2), 17–45.

Avraham, E., & Ketter, E. (2013). Marketing destinations with prolonged negative images: Towards a theoretical model. *Tourism Geographies, 15*(1), 145–64. https://doi.org/10.1080/14616688.2011.647328

Becken, S., Simmons, D. G., & Frampton, C. (2003). Energy use associated with different travel choices. *Tourism Management, 24*(3), 267–77. https://doi.org/10.1016/S0261-5177(02)00066-3

Bramwell, B., & Lane, B. (2000). Collaboration and partnerships in tourism planning. In B. Bramwell & B. Lane (Eds.), *Tourism collaboration and partnerships: Politics, practice and sustainability* (pp. 1–19). Clevedon: Channel View.

Brockington, D. (2002). *Fortress conservation: The preservation of Mkomazi Game Reserve, Tanzania*. London: James Currey.

Buzinde, C. N., Kalavar, J. M., & Melubo, K. (2014). Tourism and community well-being: The case of the Maasai in Tanzania. *Annals of Tourism Research, 44*, 20–35. https://doi.org/10.1016/j.annals.2013.08.010

Cater, E. (2006). Ecotourism as a western construct. *Journal of Ecotourism, 5*(1–2), 23–39. https://doi.org/10.1080/14724040608668445

Fletcher, R. (2014). *Romancing the wild: Cultural dimensions of ecotourism*. Durham, NC: Duke University Press.

Goldman, M. J. (2011). Strangers in their own land: Maasai and wildlife conservation in Northern Tanzania. *Conservation and Society, 9*(1), 65–79.

Grančay, M. (2020). Tourism sector in New Zealand–demand-side measures are necessary. *New Zealand Economic Papers*, 1–4, DOI: 10.1080/00779954.2020.1844787

Higgins-Desbiolles, F. (2020). Socialising tourism for social and ecological justice after Covid-19. *Tourism Geographies, 22*(3), 610–23. DOI: 10.1080/14616688.2020.1757748

Honey, M. (1999). *Ecotourism and sustainable development: Who owns paradise?* Washington, DC: Island Press.

Kenya Government (2020, June). Impact of COVID-19 on tourism in Kenya: The measures taken and the recovery pathways. *Ministry of Tourism and Wildlife Research Report*. Retrieved 30 October 2020, from https://www.tourism.go.ke.

Kiwango, W. A., Komakech, H. C., Tarimo, T. M., & Martz, L. (2015). Decentralized environmental governance: A reflection on its role in shaping wildlife management areas in Tanzania. *Tropical Conservation Science, 8*(4), 1080–97. https://doi.org/10.1177/194008291500800415

Lindsey, P., Allan, J., Brehony, P., Dickman, A., Robson, A., Begg, C.,... & Tyrrell, P. (2020). Conserving Africa's wildlife and wildlands through the COVID-19 crisis and beyond. *Nature Ecology & Evolution, 4*(10), 1300–10.

MacKenzie, C. A., Salerno, J., Hartter, J., Chapman, C. A., Reyna, R., Tumusiime, D. M., & Drake, M. (2017). Changing perceptions of protected area benefits and problems around Kibale National Park, Uganda. *Journal of Environmental Management, 200*, 217–28. doi:10.1016/j.jenvman.2017.05.078

McCabe, J., Perkin, S., & Schofield, C. (1992). Can conservation and development be coupled among pastoral people? An examination of the Maasai of the Ngorongoro conservation area, Tanzania. *Human Organization, 51*(4), 353–66.

Melubo, K. (2020). Is there room for domestic tourism in Africa? The case of Tanzania. *Journal of Ecotourism, 19*(3), 248–65. https://doi.org/10.1080/14724049.2019.1689987

Melubo, K., & Carr, A. (2019). Developing indigenous tourism in the *bomas*: Critiquing issues from within the Maasai community in Tanzania. *Journal of Heritage Tourism, 14*(3), 219–32. https://doi.org/10.1080/1743873X.2018.1533557

Melubo, K., & Lovelock, B. (2018). Reframing corporate social responsibility from the Tanzanian tourism industry: The vision of foreign and local tourism companies. *Tourism Planning & Development, 15*(6), 672–91. https://doi.org/10.1080/21568316.2017.1395357

Melubo, K., & Lovelock, B. (2019). Living inside a UNESCO world heritage site: The perspective of the Maasai community in Tanzania. *Tourism Planning & Development, 16*(2), 197–216. https://doi.org/10.1080/21568316.2018.1561505

Ministry of Natural Resources and Tourism (2017). *Environmental and social management framework for the resilient natural resources management for tourism and growth project*. Dodoma, Tanzania.

Ministry of Natural Resources and Tourism (2018a). *Tourism statistical bulletin*. Tourism Division. Dodoma, Tanzania.

Ministry of Natural Resources and Tourism (2018b). *Allocation of hunting blocks through auctioning (draft)*. Dodoma, Tanzania.

Ministry of Natural Resources and Tourism (2018c).*Taskforce report on recommendations for improvement of the tourist hunting industry: Review of the wildlife conservation act* (2009) *and tourist hunting regulations*. Dodoma, Tanzania.

Ministry of Natural Resources and Tourism (2019). *The multiple land use model of Ngorongoro conservation area: Achievements and lessons learnt, challenges and options for the future*. Dodoma, Tanzania.

Monshausen, A., Tremel, C., Plüss, C., Koschwitz, G., & Lukow, M. (2016). *Transforming tourism: The 2030 agenda for sustainable development*. Berlin: Bread for the World—Protestant Development Service. Retrieved 2 January 2021, from http://www.tourism-watch.de/files/2030_agenda_internet_en_0.pdf

Morupisi, P., & Mokgalo, L. (2017). Domestic tourism challenges in Botswana: A stakeholders' perspective. *Cogent Social Sciences, 3*(1), 1298171. https://doi.org/10.1080/23311886.2017.1298171

Nelson, R. H. (2003). Environmental colonialism: "saving" Africa from Africans. *The Independent Review, 8*(1), 65–86.

Neumann, R. P. (1998). *Imposing wilderness: Struggles over livelihood and nature preservation in Africa*. Berkeley: University of California Press.

Nyaupane, G. P., Paris, C. M., & Li, X. R. (2020). Introduction: Special issue on domestic tourism in Asia. *Tourism Review International, 24*(1), 1–4. https://doi.org/10.3727/154427220X15845838896305

Okello, M. M., Kenana, L., & Kieti, D. (2012). Factors influencing domestic tourism for urban and semiurban populations around Nairobi National Park, Kenya. *Tourism Analysis, 17*(1), 79–89. 99 total citations on Dimensions. https://doi.org/10.3727/108354212X13330406124214

Peters, A., & Higgins-Desbiolles, F. (2012). De-marginalising tourism research: Indigenous Australians as tourists. *Journal of Hospitality and Tourism Management, 19*(1), 76–84. https://doi.org/10.1017/jht.2012.7

Saarinen, J. (2009). Conclusion and critical issues in tourism and sustainability in Southern Africa. In J. Saarinen, F. O. Becker, H. Manwa, & D. Wilson (Eds.), *Sustainable tourism in southern Africa: Local communities and natural resources in transition* (pp. 269–86). Clevedon: Channel View Publications.

Salazar, N. B. (2010). *Envisioning Eden: Mobilizing imaginaries in tourism and beyond.* New York: Berghahn.

Scheyvens, R. (2007). Poor cousins no more: Valuing the development potential of domestic and diaspora tourism. *Progress in Development Studies, 7*(4), 307–25. https://doi.org/10.1177/146499340700700403

Scheyvens, R., & Biddulph, R. (2018). Inclusive tourism development. *Tourism Geographies, 20*(4), 589–609. https://doi.org/10.1080/14616688.2017.1381985

Scheyvens, R., & Hughes, E. (2019). Can tourism help to "end poverty in all its forms everywhere"? The challenge of tourism addressing SDG1. *Journal of Sustainable Tourism, 27*(7), 1061–79. https://doi.org/10.1080/09669582.2018.1551404

Shah, K., & Gupta, V. (2000). *Tourism, the poor and other stakeholders: Experience in Asia.* London: Overseas Development Institute.

Shultis, J., & Heffner, S. (2016). Hegemonic and emerging concepts of conservation: A critical examination of barriers to incorporating Indigenous perspectives in protected area conservation policies and practice. *Journal of Sustainable Tourism, 24*(8–9), 1227–42. https://doi.org/10.1080/09669582.2016.1158827

Sindiga, I. (1996). Domestic tourism in Kenya. *Annals of Tourism Research, 23*(1), 19–31. https://doi.org/10.1016/0160-7383(95)00040-2

Slocum, S. L., & Backman, K. F. (2011). Understanding government capacity in tourism development as a poverty alleviation tool: A case study of Tanzanian policy-makers. *Tourism Planning & Development, 8*(3), 281–96. https://doi.org/10.1080/21568316.2011.591157

Snyder, K. A., & Sulle, E. B. (2011). Tourism in Maasai communities: A chance to improve livelihoods? *Journal of Sustainable Tourism, 19*(8), 935–51. https://doi.org/10.1080/09669582.2011.579617

Spenceley. A. (2008). *Assessment of the development of non-consumptive tourism activities under Tanzania wildlife management authority* (TAWA). Ministry of Natural Resources and Tourism. Morogoro, Tanzania.

Spenceley. A. (2020, June 23). COVID-19 & protected area tourism data analysis. Retrieved 22 January 2021, from https://trade4devnews.enhancedif.org/en/news/covid-19-and-tourism-africas-protected-areas-impacts-and-recovery-needs

Stone, L. S., & Nyaupane, G. P. (2018). The tourist gaze: Domestic versus international tourists. *Journal of Travel Research, 58*(5), 877–91. https://doi.org/10.1177/0047287518781890

Stone, L. S., & Nyaupane, G. P. (2020). Local residents' pride, tourists' playground: The misrepresentation and exclusion of local residents in tourism. *Current Issues in Tourism, 23*(11), 1426–42. https://doi.org/10.1080/13683500.2019.1615870

Stone, L. S., & Stone, T. M. (2011). Community-based tourism enterprises: Challenges and prospects for community participation; Khama Rhino sanctuary trust, Botswana. *Journal of Sustainable Tourism, 19*(1), 97–114. https://doi.org/10.1080/09669582.2010.508527

Tanzania Wildlife Management Authority (2018). *Guidelines for the allocation of hunting blocks through auctioning.* Retrieved 14 November 2020, from https://www.tawa.go.tz/fileadmin/user_upload/Guideline-HuntingBlocks_Oct_2019.pdf

Tanzania Wildlife Management Authority (TAWA) (2020). About us. Retrieved 6 January 2021, from https://www.tawa.go.tz/conservation/protected-areas/wildlife-management-areas/

Tosun, C. (2006). Expected nature of community participation in tourism development. *Tourism Management, 27*(3), 493–504. https://doi.org/10.1016/j.tourman.2004.12.004

United Republic of Tanzania [URT] (2008). *Review of the tourist hunting industry in Tanzania.* Ministry of Natural Resources and Tourism. Dodoma, Tanzania.

World Bank (2015). *Sixth Tanzania economic update: The elephant in the room: Unlocking the potential of the tourism industry for Tanzanians.* Retrieved 23 January 2021, from https://www.worldbank.org/en/country/tanzania/publication/tanzania-economic-update-increasing-tourism-for-economic-growth

World Travel and Tourism Council (2018). Domestic tourism: Importance and economic impact. Retrieved 15 December 2020, from https://www.wttc.org/-/media/files/reports/2018/domestic-tourism--importance--economic-impact-dec-18.pdf

Section II

Socialising tourism as rethinking ideology

5 Tourism, COVID-19 and crisis

The case for a radical turn

Raoul V. Bianchi

Introduction

In recent decades, such has been the unfaltering belief in the resilience of tourism as an export sector that in January 2020 the Secretary-General of the UNWTO predicted 3%–4% growth for the coming year (UNWTO, 2020a). The rapid spread of COVID-19 quickly brought that optimism to an abrupt halt. The global scope and magnitude of the pandemic led swiftly to border closures and government-imposed national "lock-downs", resulting in a 65% fall in tourism in the first half of 2020 and concomitant decline of US$460 billion in export revenues (UNWTO, 2020b, p. 1). Depending on the continued timescale and severity of the pandemic, together with the nature and efficacy of political interventions to both stem its spread and mitigate the economic fallout, it is estimated that international tourism receipts could fall by up to US$1.2 trillion with a resultant loss of 100–120 million jobs (Richter, 2020).

The pandemic has also revealed the degree to which tourism and its associated mobility flows are deeply woven into the unsustainable processes and interconnected architecture of international tourism as well as demonstrated its capacity to act as a vector for the transmission of the disease itself (Oltermann, 2020). While the urgency with which the UNWTO (2020c) has sought to address the economic repercussions of pandemic suppression measures on tourism is perhaps understandable, the hasty reopening of many destinations in the summer of 2020 undoubtedly contributed to the resurgence of infection rates across Europe and will likely forestall any immediate recovery. Nevertheless, a number of analysts have interpreted the crisis as an unprecedented opportunity to "critically reconsider tourism's growth trajectory" (Gössling, et al., 2020, p. 13) and to reimagine a post-pandemic tourism recovery within more equitable, environmentally sustainable and regenerative frameworks (Cave & Dredge, 2020; Higgins-Desbiolles, 2020; Ioannides & Gyimóthy, 2020).

The UNWTO (2020d) meanwhile has called for the tourism industries to "grow back better". However, the rebuilding of tourism, for the most part, centres on modest health-related innovations and tourist behavioural shifts,

DOI: 10.4324/9781003164616-5

together with nods to the recipes of "inclusive" and "sustainable" growth, as outlined in the UNWTO's 2030 Sustainable Development Agenda (see UNWTO, 2017). Immediate and comprehensive measures are needed in the form of aid, debt relief and other mitigating measures to prevent a humanitarian catastrophe and permanent loss of productive capacity in low-income tourism destinations. But to ignore the structural inequalities and systemic processes of exploitation that have been characteristic of growth-led, corporate, capitalist tourism development in recent decades, would signal to powerful political and commercial interests that a serious challenge to their profit-making is unlikely to materialise. Alternatively, to delay radical reforms in the interests of restoring tourism growth in the short term would likely render attempts to address such concerns that much harder to achieve once developers have secured the capture of lucrative "tourism resources" and well-financed tourism business interests have reinforced their superior market position.

Many of the commentaries and responses thus far have failed to fully consider and critique the nature of the political and economic forces that have driven the growth and development of international and regional tourism and how these will continue to shape struggles to restructure tourism business models along more "sustainable" lines in its aftermath. Accordingly, any kind of reflection on possible alternatives to growth-led, corporate-managed, extractive tourism development, together with the necessary forms of collective action needed to socialise tourism business models in line with distributed and equitable forms of ownership and control, must address the contested class relations of power and forces of accumulation that fuel tourism capital accumulation.

COVID-19, tourism and the political economy of disruption

That this constitutes a profound crisis and major turning point for the global economy and tourism is beyond doubt, one whose economic repercussions are global in scale and likely to be of greater severity than the 1930s depression (IMF, 2020a). Whereas the 2008 financial crisis constituted a crisis of financialised capitalism, leading to a sharp contraction of liquidity, the unique nature of the widespread economic repercussions brought about by the COVID-19 pandemic cannot be alleviated by the usual recipe of macroeconomic interventions, thus increasing the likelihood of "a fundamental shift in the very nature of the global economy" (Milanovic, 2020). Given the uniquely face-to-face nature of the tourism and hospitality industries and their dependence on cross-border mobility, these industries have experienced an unprecedented contraction in consumer demand and supply because of the COVID-19 pandemic. The value produced by tourism is constituted through a myriad of interconnected firms and service suppliers of different scale and geographical reach, through which diverse capitals flow and firms coexist, from corporate-managed global commodity chains to informal traders (Gibson, 2009). Such complexity renders the intricate cross-border linkages and local networks of suppliers, all of whom are bound through their dependence upon the demand generated by tourists, at increased risk of a sustained downturn and potential collapse.

Tourism supply is by its very nature perishable and unsold inventory cannot be resold at a later date. Restrictions on overseas travel have thus led to an increased emphasis on domestic tourism, which accounts for around 75% of tourism in OECD countries, and is expected to recover more quickly (OECD, 2020). However, in export-oriented, low-income destinations, domestic tourist markets cannot be easily substituted for international tourists nor can tourism and resort infrastructures be easily repurposed for alternative economic use, as demonstrated in the Maldives where a majority of hotels and resorts remained shut until mid-July (Liang-Pholsena, 2020).

Notwithstanding the contribution made by tourism to export earnings, employment and improved living standards in many countries, the pandemic has shattered "the myth of 'catch up development' and 'perpetual growth'" (Everingham & Chassagne, 2020, p. 556). As noted in a recent report by the UN Human Rights Council between 1980 and 2016, 27% of global real income growth was captured by the top 1% of the world's population (Alston, 2020, p. 15). Not only do the financial benefits of economic growth accrue unevenly, the construction of tourism resorts and infrastructure in many growth "hotspots" in the Global South, and indeed elsewhere, often result in significant disruption to local economies, land dispossessions and environmental pressures (Diagne, 2004; Mittal & Fraser, 2018). In Africa, despite growth in international tourist arrivals (6% per year) and tourism export revenues (9% per year) between 1995 and 2014, poverty, unemployment and inequality remain high (UNCTAD, 2017, p. 15). Many destinations thus remain trapped by the competitive logics of export-led tourism which, in turn, exacerbates the unequal patterns of resource extraction (including atmospheric pollution from air travel by tourists), drives up local prices, increases inequalities and sustains precarious work.

Although it is too early for a comprehensive analysis of the industrial transformations and power coordinates of the international and regional tourism political economy that will emerge in the pandemic's wake, the social and economic distribution of the pandemic's impact on tourism firms and destinations worldwide underlines the coercive trade relations and inequalities of power and wealth that govern economic relations between the Global North and South. Salient among the inequalities and economic vulnerabilities exposed by the pandemic is the very "need" to trade (in tourism), particularly in low-income states with few economic alternatives.

Only three years ago, an UNCTAD (2017, p. 30) report on tourism in Africa identified tourism revenues as being "more stable and resilient to shocks than most other external and trade flows". However, the pandemic has laid bare long-standing inequalities and vulnerabilities in the export-led tourism development strategies pursued by "developing countries" under the aegis of Western development "experts" and agencies since the 1960s (see Wood, 1979, p. 274). Although tourism-related stocks of foreign direct investment (FDI) remain low by comparison with other sectors, a shortage of domestic capital and accumulated debt burdens – in themselves indicative of systemic trade inequalities (see Hickel, 2017, pp. 174–81) – has left many destinations reliant upon inward FDI flows to finance hotel and resort development. The lack of economic resilience is illustrated in the fact that between January and June 2020, inward FDI into tourism projects around the world fell from 373 to 141 while total capital expenditure fell by 73.3% from US$34.8 billion to US$9.3 billion (Dettoni & Conway, 2020).

These innate vulnerabilities have been most acutely felt in small island developing states (SIDS) given their extremely high dependence on tourism export revenues, which amount to approximately US$30 billion per year, high levels of external debt (72.4% of their GDP) and low levels of foreign reserves (Coke-Hamilton, 2020). Repayments on public external debt are estimated to reach nearly US$3.4 trillion in 2020 and 2021 (UNCTAD, 2020). Many indebted states, including Belize where export revenues have fallen by half due to the collapse of tourism, will be hard-pressed to finance a transition towards a decarbonised green economy without international financial assistance and debt relief (Farand, 2020). While these are eye-watering figures, consider for a moment that the world's advanced economies have thus far spent an estimated US$12 trillion on stimulus packages, or that the total value of capital in off-shore tax havens is currently estimated at approximately US$32 trillion (Christensen, 2020, p. 28). By way of comparison, the International Labour Organisation (ILO) estimates that US$45 billion would help mitigate the worst effects of job losses in low-income states (Achcar, 2020, p. 5).

The impact has also been severe on such countries as Costa Rica, Greece, Morocco, Portugal, and Thailand, with tourism revenue losses exceeding 3% of GDP (IMF, 2020b). Destinations in southern Europe, in which over half of all European tourism enterprises are concentrated (Eurostat, 2020), have also been particularly hard hit (Pasquale, 2020). Prior to the pandemic, the economic vulnerability of southern European economies had already been accentuated by the cumulative impacts of deindustrialisation, neoliberal reforms and European monetary union. These stimulated a surge of tourism and real estate investment in housing and other property assets. In 2015 alone, 60% of hotel investments in Barcelona were realised through investment funds (Ajuntament de Barcelona, 2017, p. 28). Further, the EU-led bailout and austerity programme implemented in the aftermath of the financial crisis in 2007–8, ostensibly to prevent default in southern eurozone economies, further eroded their economic resilience as housing was transformed into short-term property assets and digital platforms fuelled the unsustainable growth of urban tourism in such cities as Lisbon and Barcelona (Cocola Gant & Gago, 2019).

Without sustained support, it is estimated that up to 50% of small to medium-sized tourism enterprises in the OECD area are at risk of bankruptcy (OECD, 2020, p. 12). However, it is the millions of precariously employed, low-paid tourism and hospitality workers, many of whom work in the informal economy in the Global South and for whom little or no safety net exists, who have borne the brunt of the pandemic's economic shock (Baum et al., 2020). In the global cruise industry, a sector noted for its high rates of labour exploitation despite revenues of nearly US$50 billion in 2018 (Crockett, 2020), the pandemic has left a significant number of low-paid cruise ship workers stranded at sea and unable to return home (Higgins-Desbiolles, 2020). Notwithstanding access to state furlough schemes, tourism and hospitality workers in wealthy states have also not been spared. In the United States, 98% of the members of the Unite Here trade union (representing hotel, casino and food service workers) had lost their jobs by early April (Bandler, 2020). Meanwhile, in Europe almost the entire 12 million strong hospitality workforce has either been furloughed or been made redundant (Bragason, 2020).

Conversely, further to being able to access state financial support under the auspices of the European Union's "Temporary Framework" (CEC, 2020), many large-scale European-based tourism and hotel corporations have been able to benefit directly from state financial support. Hotel chains and resort investment companies with substantial real estate holdings may also be able to sell real estate assets to release liquidity. German-based tour operator TUI secured €3 billion in financing via state-owned bank KfW, although this failed to prevent them from shedding 8,000 jobs and freezing payments to hoteliers (Hancock, 2020a). Meanwhile smaller, independent travel firms have struggled to survive (e.g. *Hays Travel*).

Although occupancy rates have plummeted during the early pandemic, some estimates suggest that global hotel supply would contract by a mere 2% largely thanks to the ability of large US-based corporations to access state aid and lines of credit (Sperance, 2020). In contrast, the many thousands of small to medium-sized firms, which make up around 80% of the global tourism industries, have struggled to access emergency government assistance and thus remain particularly vulnerable to collapse (UNWTO, 2020b). These two factors combined are likely to drive an increase in mergers and acquisitions as smaller hotel companies and groups have little choice but to be absorbed by larger entities (Hotel News Resource, 2020).

Nor are the increasingly dominant corporate digital platforms immune. Expedia announced 3,000 job losses in February 2020 (Hancock, 2020b), and despite revenues of US$1.5 billion in the third quarter of the year, it has accumulated losses of US$220 million (Eckstein, 2020). Nevertheless, the dominance of corporate, digital tech companies is likely to accelerate and intensify, not least Airbnb, which intends to raise US$2.5 billion from its recent flotation (Karnevali & Kruppa, 2020), and Google, which has become a comprehensive meta-search platform for all manner of travel-related services (Airport Technology, 2020). Amongst the hardest hit of all travel- and tourism-related sectors are the airlines and cruise industries. According to more recent estimates from the International Air Transport Association (IATA), global airlines face revenue losses of over US$84 billion and 25 million jobs out of a total of 65.5 million employed in the global air transport industries (IATA, 2020a, 2020b). However, thanks in part to aggressive lobbying and their strategic importance, many of the world's major airlines have managed to secure vital state aid to stay afloat (Barratt, 2020). Indeed, while US lobby "Airlines for America" has consistently campaigned against aviation taxes and regulation on behalf of US airlines, passenger carriers have been amongst the major beneficiaries of state financial support (US$46 billion) made available to the aviation sector under the US Coronavirus Aid, Relief and Security Act (Brenner, 2020).

However, in a sign of the schism between the interests of transnational corporations and states characteristic of the era of neoliberal globalisation, the three nominally US companies that account for 75% of the global cruise market (Crockett, 2020) were excluded from the US$500 billion state support programme by virtue of being incorporated in off-shore registries ("flags of convenience") which enables them to minimise US tax liabilities and circumvent other "restrictive" labour and environmental standards (Zeballos-Roig, 2020). There are signs too that the pandemic may hasten the geo-economic power shifts and growing distinctiveness of regional political economies that were already apparent prior to the outbreak. In this regard, the pandemic has accelerated the move of the centres of tourism capital accumulation towards East Asia and other "emerging" economies as non-Western sovereign wealth funds move to purchase equity stakes in a range of aviation, hotel, cruise and entertainment companies (Massoudi et al., 2020).

From crisis to renewal?

Crises often serve to catalyse sharp transitions towards new political-economic orders. The degree to which the current rupture signals "the decline of mainstream business formats" and an unprecedented opportunity to adopt a "more sustainable path" in tourism (Ioannides & Gyimóthy, 2020) remains uncertain, and will depend upon various factors internal and external to destinations themselves. The COVID-19 pandemic does indeed have all the attributes of just such a crisis. However, while some have hailed the pandemic as the death knell for neoliberal capitalism, crises (even pandemics) are not simply natural events, but rather "socially constructed and highly political" (Gamble, 2009, p. 38). Given that ample evidence of unsustainable and exploitative tourism practices has existed for at least 30 years, if not longer, the question is: why would governments, developers and corporations choose to make substantive changes to capitalist business practices now?

There are indications that citizens and local governments are increasingly prepared to challenge the untrammelled growth of tourism in such cities as Barcelona (Blanco-Romero et al., 2018) and Venice (Croce, 2020). In Key West, Florida, citizens have recently also voted to prohibit large cruise ships from docking and limiting the number of disembarking passengers (Filosa, 2020). However, the struggle to define the precise parameters of new tourism business models and to catalyse a transformation towards a more equitable, just and sustainable tourism political economy has only just begun. Aside from these examples, thus far there are few signs that the pandemic will significantly lead to a major challenge to the legitimacy of the current political and economic order. If previous crises are any indication, such disruptions often provide an opportunity for corporate and state actors to gain or reinforce their access and control over common pool resources which may then be exploited as profitable tourism assets (Ratnayake & Hapugoda, 2017).

There are already signs that large, well-capitalised firms are positioned to exploit the crisis to accelerate and consolidate their market dominance. Accor, the world's fifth largest hotel group by turnover, continues to expand its portfolio of hotels, while Radisson has announced plans to expand their property portfolio in Africa by 50% (Atkins, 2020). The pandemic has also sharpened certain mobility injustices as, for example, wealthy urbanites in Europe retreated to second homes in the countryside while essential workers, including immigrant health, transport and delivery workers, were thrust into the front lines of the pandemic. Meanwhile demand for luxury tourism, yachts and private jets has soared as the wealthy seek out remote, exclusive, sanitary environments (Rodrigues, 2020).

One of the most marked effects of the pandemic has been to highlight just how vulnerable three decades of hyper-globalisation and market-led growth has rendered even advanced capitalist economies (Fouskas & Roy-Mukherjee, 2020). However, while many governments have stepped in to prevent the collapse of businesses and mitigate the effects of spiralling unemployment (IMF, 2020c), to conclude that we are witnessing the demise of financialised neoliberal capitalism would be premature. Moreover, the conditionalities attached to state aid and support for industry vary from country to country. State financial support (€10 billion) for Air France-Lufthansa was conditional on a commitment to halving emissions by 2030, the elimination of certain domestic airline routes and suspension of divided payments. As modest as these provisions are, no such conditions accompanied the £300 million British government loan to British Airways nor did the UK government seek to find ways to avoid the 10,000 job losses announced by the airline in August. It was only thanks to lobbying from trade unions that the airline reversed its decision to rehire remaining employees on reduced pay and worse contractual conditions (Georgiadis, 2020). Thus, while thousands of small businesses struggle to access state aid and/or lines of credit, large tourism-hospitality-leisure corporations (with the exception of the cruise industry) have seen their risks socialised and the upwards (re)distribution of wealth underwritten by the state (i.e. tax revenues). If anything, it would have been much simpler, and arguably fairer, to transfer much of the financial support straight into the pockets of workers themselves rather than channel it via corporate executives.

Beyond sustainable tourism: the politics of post-pandemic tourism development

The precise socio-political coordinates of struggle to rebuild and transform tourism in the aftermath of the pandemic will be shaped by the geographic variances of tourism and capitalist development. This includes, inter alia, the nature and character of the state, the class relations of power which shape the flows and composition of inward tourism investment and tourism divisions of labour as well as the scale and character of tourism development in different destinations. One thing is certain, however, as the pandemic subsides and destinations reopen, it will be hard for governments to resist commercial pressures to restore growth and profitability and to push back against corporate lobbying demanding the loosening of fiscal "burdens" and "restrictive" social and environmental regulations.

Democratic forms of public control over tourism capital and investment are needed in order to sever the grip of markets and capital on tourism governance and destination planning. Capital flight is a constant threat to the stability of tourism economies in the Global South. Indeed, it has often been deployed as a bargaining tool to forestall the threat of radical reform, as Jamaican Premier Michael Manley discovered in the 1970s when tourism investors and their allies in the Jamaican capitalist class engineered a managed disinvestment in response to his progressive social and economic agenda (Editors, 2020). Similar outcries from the tourism and hotel corporate executives greeted the Balearic eco-tax in 2001 and its more recent iteration (Ultima Hora, 2019).

The dependence of destinations on the growth-led logics of tourism at the expense of ecological degradation, residential displacement and socio-economic inequalities points to a series of inherent contradictions in the capitalist political economy of tourism, which cannot easily be resolved without confronting the predatory modes of profit extraction. The perpetual promise of jobs and prosperity functions as the ideological handmaiden of growth-led capitalist tourism development, while simultaneously foreclosing demands for redistribution or reducing the pressures of construction on fragile landscapes.

The complex and often opaque structure of corporate ownership across the tourism industries, the drivers of growth and systemic capital accumulation thus present a formidable obstacle to radical change. The pursuit of growth and concomitant expansion of urban development for tourism, with all the prospects for profiteering that this entails, often occurs at the expense of productive investment and acts as a substitute for innovation that may boost productivity and create the potential for a more inclusive tourism development model. While the land on which resorts and hotels are built can be subject to restrictive environmental ordinances to prevent the degradation of ecosystems, the corporate assets (hotel and resort companies) themselves are mobile and harder to regulate. Thus, any kind of paradigmatic shift in tourism will require a challenge to/dismantling of the tourism-real estate nexus that underpins the political economy of tourism in many, although by no means all, destinations worldwide. Neither are cities, whose urban property markets have increasingly proven to be lucrative investment vehicles, immune to the destabilising effects of speculative property capital.

The transition to sustainable, equitable and resilient tourism economies is thus not simply a question of fine-tuning regulation and imposing limits to tourism growth (although this may well be part of the solution). There is little use talking of "rebuilding" or indeed socialising tourism, if inequitable property relations, corporate ownership structures and exploitative labour practices are beyond critical scrutiny. The limited horizons of inclusive growth and use of technocratic digital technologies to manage "Covid-secure", post-pandemic mobilities comprise neither adequate nor indeed sufficiently democratic responses to the range of structural problems exposed by the crisis. Technocratic, data-driven "solutions" reduce the question of the sustainable rebuilding of destinations and equitable tourism economies to one of choosing the correct algorithm. This approach risks promoting a technocratic, corporate-managed sustainable tourism in the absence of democratically agreed principles and values around which to frame the intended outcomes from tourism development going forward into the post-pandemic era.

Conclusion

The precise pathways to a sustainable recovery and equitable transformation will not be linear, but will vary according to the local structure and organisation of tourism capital and its relationship to the changing tectonics of global macroeconomic forces. The transition away from a growth-led extractive model of tourism capital accumulation will not be addressed through socialised, community-based tourism business models alone, many of which have existed at the margins of the capitalist political economy of tourism for some time (see Saglio, 1979). Rather, it presupposes coordinated transnational action to control disruptive flows of capital and diminish the power of predatory profit-making at a wider level. This will also necessarily involve state-managed transitions towards decarbonised, intermodal transport systems prioritising the training and redeployment of workers from fossil fuel-dependent transport industries (see Descamps, 2020). Only in this way can the tensions between tourism growth and ecological limits, often invoked by developers to seek legitimacy for corporate-led tourism projects, be dissolved (see Boissevain & Theuma, 1998).

Tourism investments must be severed from the short-term logics of speculative real estate through the strengthening of democratic public control over tourism finance, resources and the productive capacity of destination themselves. For example, *land*, a relatively immobile resource and much sought-after commodity by financiers and tourism developers, can be brought under more equitable, distributed forms of ownership while remaining open to commercial use (see Christophers, 2019, p. 47). This will involve more than localising destination value chains or simply calling for the inclusion of marginalised voices into existing decision-making apparatuses. It implies a progressive and robust politics of intervention at all scales, in order to challenge and disempower the nexus of commercial-financial-political interests which have abetted the relentless growth of tourism and expansion of capital accumulation through land grabs, privatisations, regulatory liberalisation and real estate-driven strategies of tourism development. The re-engineering of their respective city economies away from speculative tourism and real estate-driven growth by city councils in Barcelona and Lisbon as well as support for restrictions on cruise ship docking in Key West, Florida, suggest that activist citizens, particularly where robust social movements exist, are prepared to challenge some of the most egregious consequences of untrammelled tourism growth if not the logic of profit-driven tourism economies altogether. However, the socialisation of tourism benefit structures cannot be sustained via the efforts of activist citizen coalitions to simply constrain the power of capital, but necessarily involve building alliances with those who labour in the tourism industries to democratise the economy itself.

Any challenge to the monopolistic concentrations of corporate power, speculative finance and offshore capital that flow through tourism destinations will almost certainly involve sustained resistance from capitalist business interests and perhaps the state itself. However, it should be recognised that much of the private wealth generated from tourist activity by corporate investors, landlords and speculators is in fact *collectively* produced out of a dynamic assemblage of use and exchange values while simultaneously obscuring the class character of capital accumulation (see Young & Markham, 2019). Many of those who "produce" the value that contributes to tourism revenues are often not directly involved in the "tourism industry" at all. For example, it is estimated that the fishermen in the town of Hastings, England, contribute £5 million to the local economy through their mere presence (Toynbee, 2018).

With this in mind, there is potential scope to democratise the gains from the tourism economy through cooperative forms of ownership and equitable models of tourism revenue-sharing (e.g. Mtapuri & Giampiccoli, 2020). One such model exists in the form of the Alaska Permanent Fund Dividend through which all Alaskan citizens receive an annual payment from the state (Widerquist, 2013). However, where this fund is based upon a portfolio of state-managed investments derived from oil and gas revenues, a similar annual collective "destination dividend" could be used to provide a guaranteed income to all residents of a destination regardless of whether or not they are employed in the tourism industry. A dividend of this kind could potentially help to socialise the revenues and enhance the distributive outcomes arising from tourism activities while reducing the pressures on destination ecosystems borne out of the competitive struggle to capture and exploit strategic tourism resources.

Despite scattered signs of change, the potential for socialising tourism beyond a few localised settings remains severely constrained by the current concentrations of economic power underpinning monopoly-digital capitalism and its influence on the political economy of tourism at all levels. Moreover, many of the proposed alternatives to the dominant paradigm of corporate-mass tourism remain disconnected from a more rigorous inspection of political economy and the contested class relations that shape and determine the creation of value and distributive outcomes in tourism. To paraphrase climate activist Bill McKibben (2020), progressive voices in tourism may have begun to win the argument concerning the need for an urgent transformation of tourism, but are far from winning the struggle to catalyse the transformation towards a more ecologically sustainable and democratically controlled tourism political economy.

Capitalist tourism development – in all its variants, not merely its "neoliberal" form – cannot be understood simply as a "project" consciously engineered by corporations and their political handmaidens, which can therefore be transformed through an appeal to alternative *values* alone. Neither can a radical and just transition away from corporate-managed and extractive models of tourism development and the disciplinary constraints imposed on progressive governance by financial markets focus solely on questions of restraining tourism growth (much of which, in any case, derives from speculative financial and digital transactions rather than productive investment) or advancing its putative opposite – degrowth. Rather, it is necessary to dismantle the grip of unrelenting logic of profit-making and destructive competition that leads to the massively uneven concentrations of wealth and environmental degradation through the constant expansion of tourism. Such a strategy must be built on a platform of the democratic civic and public management of tourism's material growth within agreed limits, in conjunction with the socialisation of the assets and cooperative use of the resources upon which tourism and associated local livelihoods depend. The COVID-19 pandemic has visibly revealed the manifold injustices and vulnerabilities in global capitalism as well as highlighted much of the hubris upon which the ceaseless expansion of growth and profit-making driving mass travel and tourism has been built. Now is not the time to surrender to the seductive deceit of pragmatism – it is time for radical change.

Acknowledgements

The author would like to thank Frans de Man, Director of the Retour Foundation, Netherlands, for his comments and input into this chapter. All interpretations and opinions are nevertheless the author's own.

References

Achcar, G. (2020). The great lockdown hits the Third World hard. *Le Monde Diplomatique*, November. Retrieved 6 November 2020, from https://mondediplo.com/2020/11/04covid-third-world

Airport Technology (2020). Google's monopoly of the online travel space may intensify. 17 April 2020. Retrieved 6 November 2020, from https://www.airport-technology.com/comment/google-online-travel-space-COVID-19/

Ajuntament de Barcelona (2017). *Turisme 2020 Barcelona: Pla Estrategic*. Barcelona: Direcció de Turisme.

Alston, P. (2020). *Report on the parlous state of poverty eradication*. Report of the Special Rapporteur on extreme poverty and human rights. Geneva: UNHCR. Retrieved 22 October 2020, from https://www.ohchr.org/EN/Issues/Poverty/Pages/parlous.aspx

Atkins, W. (2020). Radisson bets on African growth with new hotels. *fDi Intelligence*, 9 September. Retrieved 12 October 2020, from https://www.fdiintelligence.com/article/78380

Bandler (2020). Inside the union where coronavirus put 98% of members out of work. *ProPublica*, 9 April. Retrieved 13 November 2020, from https://www.propublica.org/article/inside-the-union-where-coronavirus-puts-98--of-members-out-of-work

Barratt, L. (2020). Documents reveal airline industry plan for tax breaks, subsidies and voucher refunds. *Unearthed*, 7 April. Retrieved 6 November 2020, from https://unearthed.greenpeace.org/2020/04/07/coronavirus-airlines-lobby-for-tax-breaks-subsidies-vouchers-passenger-refunds/

Baum, T., Mooney, S. K. K., Robinson, R. N. S., & Solnet, D. (2020). COVID-19's impact on the hospitality workforce – new crisis or amplification of the norm? *International Journal of Contemporary Hospitality Management, 32*(9), 2813–29. https://doi.org/10.1108/IJCHM-04-2020-0314

Blanco-Romero, A., Blázquez-Salom, M., & Cànoves, G. (2018). Barcelona, housing rent bubble in a tourist city. Social responses and local policies. *Sustainability, 10*(6), 1–18. https://doi.org/10.3390/su10062043

Boissevain, J., & Theuma, N. (1998). Contested space: Planner, s, tourists and environmentalists in Malta. In S. Abram & Waldren, J. (Eds.), *Anthropological perspectives on local development* (pp. 96–119). London: Routledge.

Bragason, K. (2020). A sector on ice – tourism in Europe amid COVID-19. *European Federation of Food, Agriculture and Tourism Trade Unions*. 23 April. Retrieved 13 November 2020, from https://effat.org/in-the-spotlight/a-sector-on-ice-tourism-in-europe-amid-COVID-19/

Brenner, R. (2020, May–June). Escalating plunder. *New Left Review, 123*, 5–22.

Cave, J., & Dredge, D. (2020). Regenerative tourism needs diverse economic practices, *Tourism Geographies, 22*(3), 503–13. https://doi.org/10.1080/14616688.2020.1768434

CEC (2020). *Temporary Framework for State aid measures to support the economy in the current COVID-19 outbreak*. 19 March, C(2020) 1863 final. Brussels: Commission of the European Union. Retrieved 20 November 2020, from https://ec.europa.eu/competition/state_aid/what_is_new/covid_19.html

Christensen, J. (2020, March–April). Who's the thief? *New Internationalist*, 524, 26–28.

Christophers, B. (2019). *The new enclosure: The appropriation of public land in Neoliberal Britain*. London: Verso.

Cocola Gant, A., & Gago, A. (2019). Airbnb, buy-to-let investment and tourism-driven displacement: A case study in Lisbon, *Environment and Planning A: Economy and Space*, Published online, 19 August 2019. https://doi.org/10.1177/2F0308518X19869012

Coke-Hamilton, P. (2020). *Impact of COVID-19 on tourism in small island developing states*. 24 April. Geneva: United Nations Conference on Trade and Development. Retrieved 22 October 2020, from https://unctad.org/news/impact-COVID-19-tourism-small-island-developing-states

Croce, S. (2020). Venezia. Dall'overtourism al turismo sostenible. *Ytali*, 24 April. Retrieved 29 October 2020, from https://ytali.com/2020/04/24/venezia-dallovertourism-al-turismo-sostenibile/

Crockett, Z. (2020). The economics of cruise ships. *The Hustle*, 15 March. Retrieved 13 November 2020, from https://thehustle.co/the-economics-of-cruise-ships/

Descamps, P. (2020). No going back to business as usual. *Le Monde Diplomatique*, July. Retrieved 5 December 2020, from https://mondediplo.com/2020/07/09aviation

Dettoni, J., & Conway, J. (2020). Covid-19 upends tourism sector after record run. *fDi Intelligence* (*Tourism Investment Report 2020*).17 September. Retrieved 15 October 2020, from https://www.fdiintelligence.com/article/78690

Diagne, A. (2004). Tourism development and its impacts the Senegalese Petite Côte: A geographical case study in centre-periphery relations, *Tourism Geographies, 6*(4), 472–92. https://doi.org/10.1080/1461668042000280246

Eckstein, N. (2020). Mass tourism will be roaring back by summer, Says Expedia CEO. *Bloomberg*, 16 November. Retrieved 25 November 2020, from https://www-bloomberg-com.cdn.ampproject.org/c/s/www.bloomberg.com/amp/news/articles/2020-11-16/mass-tourism-will-be-roaring-back-by-summer-says-expedia-ceo

Editors (2020). We won the battle they won the war. *Jacobin*, No 36, 19 February. Retrieved 27 November 2020, from https://www.jacobinmag.com/2020/02/we-won-the-battle-they-won-the-war

Eurostat (2020). Tourism vital to employment in several Member States. 15 April. Retrieved 6 November 2020, from https://ec.europa.eu/eurostat/web/products-eurostat-news/-/DDN-20200415-1?inheritRedirect=true&redirect=%2Feurostat%2Fnews%2Fwhats-new

Everingham, P., & Chassagne, N. (2020). Post COVID-19 ecological and social reset: Moving away from capitalist growth models towards tourism as Buen Vivir. *Tourism Geographies, 22*(3), 555–66. https://doi.org/10.1080/14616688.2020.1762119

Farand, C. (2020). Ballooning debt cripples poor countries' hopes of green recovery from Covid. *Climate Home News*, 15 October. Retrieved 23 October 2020, from https://www.climatechangenews.com/2020/10/15/ballooning-debt-cripples-poor-countries-hopes-green-recovery-covid/

Filosa, G. (2020). Key West voters put limits on cruise ships but a legal battle looms. *Miami Herald*, 3 November. Retrieved 20 November 2020, from https://www.miamiherald.com/article246615598.html

Fouskas, V., & Roy-Mukherjee, S. (2020). Corona crisis: There Is No Alternative (TINA) – to Socialism, this Time. *Brave New Europe*, 25 March. Retrieved 26 November 2020, from https://braveneweurope.com/vassilis-k-fouskas-shampa-roy-mukherjee-there-is-no-alternative-tina-to-socialism-this-time

Gamble, A. (2009). *Spectre at the feast: Capitalist crisis and the politics of recession*. Basingstoke: Palgrave Macmillan.

Georgiadis, P. (2020). BA to drop controversial 'fire and rehire' plan for thousands of staff. *Financial Times*, 16 September. Retrieved 26 November 2020, from https://www.ft.com/content/455bc880-9d86-42fd-9293-530db8f1262e

Gibson, C. (2009). Geographies of tourism: Critical research on capitalism and local livelihoods. *Progress in Human Geography, 33*(4), 527–34. https://doi.org/10.1177/0309132508099797

Gössling, S., Scott, D., & Hall, C. M. (2020). Pandemics, tourism and global change: A rapid assessment of COVID-19. *Journal of Sustainable Tourism, 29*(1), 1–20. https://doi.org/10.1080/09669582.2020.1758708

Hancock, A. (2020a). Tui secures €1.8bn loan to navigate coronavirus crisis. *Financial Times*, 27 March. Retrieved 26 November 2020, from https://www.ft.com/content/6f17c632-abe9-4b6a-adba-1d05db3cda99

Hancock, A. (2020b). Expedia cuts 3,000 jobs in major restructuring. *Financial Times*. Retrieved 13 November 2020, from https://www.ft.com/content/26437d0e-57c2-11ea-a528-dd0f971febbc

Hickel, J. (2017). *The divide: A brief guide to global inequality and its solutions*. London: William Heinemann.

Higgins-Desbiolles, F. (2020). Socialising tourism for social and ecological justice after COVID-19. *Tourism Geographies, 22*(3), 610–23. https://doi.org/10.1080/14616688.2020.1757748

Hotel News Resource (2020). Surge in mergers and acquisitions activity likely for hotel industry post-COVID-19. 7 May. Retrieved 13 November 2020, from https://www.hotelnewsresource.com/article110549.html

IATA (2020a). Industry losses to top $84 billion in 2020. 9 June. Retrieved 6 November 2020, from https://www.iata.org/en/pressroom/pr/2020-06-09-01/

IATA (2020b). 25 million jobs at risk with airline shutdown. 7 April. Retrieved 6 November 2020, from https://www.iata.org/en/pressroom/pr/2020-04-07-02/

IMF (2020a). The great lockdown: Worst economic downturn since the great depression. *IMF Blog*, 14 April. Retrieved 14 October, from https://blogs.imf.org/2020/04/14/the-great-lockdown-worst-economic-downturn-since-the-great-depression/

IMF (2020b). *External sector report: Global imbalances and the COVID-19 crisis*. Washington, DC: International Monetary Fund.

IMF (2020c). State-owned enterprises in the time of Covid-19. *IMF Blog*, 7 May. Retrieved 14 October 2020, from https://blogs.imf.org/2020/05/07/state-owned-enterprises-in-the-time-of-covid-19/

Ioannides, D., & Gyimóthy, S. (2020). The COVID-19 crisis as an opportunity for escaping the unsustainable global tourism path. *Tourism Geographies, 22*(3), 624–32. https://doi.org/10.1080/14616688.2020.1763445

Karnevali, D., & Kruppa, M. (2020). Airbnb looks to raise up to $2.5bn in IPO. *Financial Times*, 1 December 2020. Retrieved 11 December 2020, from https://www.ft.com/content/c5450812-c45d-4833-a99f-22a390e5d3e4

Liang-Pholsena, X. (2020). When you don't have any domestic tourism to rely on: Maldives as a pandemic case study. *Skift*, 1 May. Retrieved 23 October 2020, from https://skift.com/2020/05/01/when-you-dont-have-any-domestic-tourism-to-rely-on-maldives-as-a-pandemic-case-study/

Massoudi, A., Hancock, A., & England, A. (2020). Saudi Arabia's PIF takes 8.2% stake in cruise operator Carnival. *Financial Times*, 6 April. Retrieved 13 November 2020, from https://www.ft.com/content/8dd37fe5-64d0-4524-b115-86aa5a6a7bd0

McKibben, B. (2020) 'A Bomb in the center of the climate movement': Michael Moore damages our most important goal. *Rolling Stone*, 1 May. Retrieved 10 December 2020, from https://www.rollingstone.com/politics/political-commentary/bill-mckibben-climate-movement-michael-moore-993073/

Milanovic, B. (2020). The real pandemic danger is social collapse. *Foreign Affairs*, 19 March 2020. Retrieved 23 October 2020, from https://www.foreignaffairs.com/articles/2020-03-19/real-pandemic-danger-social-collapse

Mittal, A., & Fraser, E. (2018). *Losing the Serengeti: The Maasai land that was to run forever.* Oakland, CA: Oakland Institute.

Mtapuri, O., & Giampiccoli, A. (2020). Toward a model of just tourism: A proposal. *Social Science, 9*(4), 1–19. https://doi.org/10.3390/socsci9040034

OECD (2020). *Tourism policy responses to the coronavirus (COVID-19).* 2 June 2020. Geneva: OECD. Retrieved 29 October 2020, from https://www.oecd.org/coronavirus/policy-responses/tourism-policy-responses-to-the-coronavirus-COVID-19-6466aa20/

Oltermann, P. (2020). Failures at Austrian ski resort "helped to speed up spread" of Covid-19 in Europe. *The Guardian*, 12 October 20202. Retrieved 14 October 2020, from https://www.theguardian.com/world/2020/oct/12/failures-at-austrian-ski-resort-helped-speed-up-spread-of-covid-19-in-europe

Pasquale, M. (2020). The tourists are leaving Italy. Now catastrophe looms. *CNN Travel*, 13 September. Retrieved 6 November 2020, from https://edition.cnn.com/travel/article/italy-tourism-economy-catastrophe-covid/index.html

Ratnayake, I., & Hapugoda, M. (2017). Land and tourism in post-war Sri Lanka: A critique on the political negligence in tourism. In A. Saufi, I. R. Andilolo, N. Othman & A. A. Lew (Eds.), *Balancing development and sustainability in tourism destinations* (pp. 221–31). Singapore: Springer.

Richter, F. (2020). COVID-19 could set the global tourism industry back 20 years. *World Economic Forum*, 2 September. Retrieved 22 October 2020, from https://www.weforum.org/agenda/2020/09/pandemic-covid19-tourism-sector-tourism/

Rodrigues, A. (2020). The pandemic is sanitizing the image of private travel. *Vice*, 25 September. Retrieved 20 November 2020, from https://www.vice.com/en/article/akzbkj/the-pandemic-is-sanitizing-the-image-of-private-travel

Saglio, C. (1979) Tourism for discovery: A project in Lower Casamance, Senegal. In E. de Kadt (Ed.), *Tourism. Passport to development? Perspectives on the social and cultural effects of tourism in developing countries* (pp.321–35). New York: Oxford University Press.

Sperance, C. (2020). Global hotels will lose just 2 percent of supply permanently because of Coronavirus: Report. *Skift*, 6 May. Retrieved 6 November 2020, from https://skift.com/2020/05/06/global-hotels-will-lose-just-2-percent-of-supply-permanently-because-of-coronavirus-report/

Toynbee, P. (2018). Propaganda delivered the Brexit vote but it can't land more fish. *The Guardian*, 23 April. Retrieved 27 November 2020, from https://www.theguardian.com/commentisfree/2018/apr/23/propaganda-brexit-fish-eu-britain-fishing-rights

Ultima Hora (2019). Escarrer dice que el impuesto turístico es una «aberración» que no «se destina ni al turismo ni al medioambiente». *UH Noticias*, 24 May. Retrieved 27 November 2020, from https://www.ultimahora.es/noticias/local/2019/05/24/1082867/escarrer-dice-impuesto-turistico-aberracion-destina-turismo-medioambiente.html

UNCTAD (2017). *Economic development in Africa report 2017. Tourism for transformative and inclusive growth.* New York & Geneva: United Nations.

UNCTAD (2020). COVID-19 is a matter of life and debt, global deal needed. 23 April. Retrieved 23 October 2020, from https://unctad.org/en/pages/newsdetails.aspx?OriginalVersionID=2339

UNWTO. (2017). *Tourism and the sustainable development goals – journey to 2030.* Madrid: UN World Tourism Organisation.

UNWTO (2020a). International tourism growth continues to outpace the global economy, 20 January. Retrieved 12 October 2020, from https://www.unwto.org/international-tourism-growth-continues-to-outpace-the-economy

UNWTO (2020b). *World tourism barometer.* Vol. 18, 5. Retrieved 12 October 2020, from https://www.e-unwto.org/doi/epdf/10.18111/wtobarometereng.2020.18.1.5

UNWTO (2020c). No time for timid leadership – the safe restart of tourism is possible. 18 August. Retrieved 15 October 2020, from https://www.unwto.org/news/no-time-for-timid-leadership-the-safe-restart-of-tourism-is-possible

UNWTO (2020d). UNWTO launches a call for action for tourism's COVID-19 mitigation and recovery, 1 April. Retrieved 14 October 2020, from https://www.unwto.org/news/unwto-launches-a-call-for-action-for-tourisms-covid-19-mitigation-and-recovery.

Widerquist, K. (2013). The Alaska model: A citizen's income in practice. *Open Democracy*, 23 April. *Open Democracy.* Retrieved 27 November 2020, from https://www.opendemocracy.net/en/opendemocracyuk/alaska-model-citizens-income-in-practice/

Wood, R. E. (1979). Tourism and underdevelopment in Southeast Asia. *Journal of Contemporary Asia, 9*, 274–87.

Young, F., & Markham, M. (2019). Tourism, capital and the commodification of place. *Progress in Human Geography, 44*(2), 1–21. https://doi.org/10.1177/0309132519826679

Zeballos-Roig, J. (2020). Here's why the depleted cruise line industry will be one of the biggest losers of the new $500 billion corporate bailout program. *Business Insider*, 4 April. Retrieved 13 November 2020, from https://www.businessinsider.com/cruise-lines-coronavirus-corporate-bailout-program-economy-trump-2020-4?r=US&IR=T.

6 The Dylann Roof road trip

A report on the banality of evil

Rasul A. Mowatt

The Dylann Roof road trip

Figure 6.1 License plate cover with three flags of the Confederate States of America. A similar plate cover was on Dylann Roof's Hyundai Elantra. Photo by author.

On the evening of 17 June 2015, Dylann Roof had responded to Tywanza Sanders: "I have to do this...ya'll taking over the world". Roof then proceeded to pump four shots into him, while continuing to shoot others in the Charleston, South Carolina-based church. Eight times. Four Times. Three times. A second eight times. Seven times. Eleven times. Six times. Another five times. Fired 77 rounds. Nine people in total were shot multiple times and killed at c. 9:05pm. Nine people, whose names are: Cynthia Graham Hurd, age 54; Susie J. Jackson, age 87; Ethel Lee Lance, age 70; Depayne Middleton-Doctor, age 49; pastor and South Carolina State Senator Clementa C. Pinckney, age 41; the aforementioned Tywanza Kibwe Diop Sanders, age 26; Daniel Lee Simmons Jr., age 74; Sharonda Coleman-Singleton, age 45; and Myra Singleton Quarles Thompson, age 59. Nine people

DOI: 10.4324/9781003164616-6

that lost their lives at the Emanuel African Methodist Episcopal (AME) Church (affectionately known as Mother Emanuel among its congregation) as they attended a bible study and prayer service with their killer. Only three others in the Church survived. In his confession, Roof thought he shot and killed "if [he] was gonna guess, five" people, but was surprised to learn that it was nine. Telling investigator's: "there wasn't even nine people there! Are you guys lying to me?". He chose this particular church because it was historic and drove to it after carefully going on extensive road trips to other sites. He chose churchgoers, because drug dealers would not send the right message. He chose Black people because that is who needed to be taken care of the most. And he was planning a trip to Nashville, Tennessee, afterwards, because he had "never been to Nashville". As Dylann Roof was asked during his trial where did he get his information from to justify what he did and to decide on the site for his act, and well it should be obvious to us because "it's all there on the Internet" (Cobb, 2017; Smith et al., 2020).

Besides the string of fake news and real news that appeared on the internet, there was a website entitled *The Last Rhodesian* at lastrhodesian.com. The site that had some 60 photographs contained on it and a 2,444-word manifesto broken into chapters of disposable populations: "Blacks", "Jews", "Hispanics" and "East Asians". The site, owned and operated by Roof, highlighted his notes, line of thinking and more germane to this discussion on travel, photos of other sites that he visited in a series of road trips that ultimately led up to his act of ethnic cleansing that he hoped would be a call to arms intended to "worsen race relations, increase racial tensions that would lead to a Race war" (Cobb, 2017). Roof was still modifying the site at 4:44pm on the day of killing and made note that he was at "the time of writing…in a great hurry", because after all he had to drive to the Church before it closed for the day. Roof was eventually arrested in Shelby, North Carolina the next morning at 10:44 am, some 245 miles away from Charleston and the Church on his way to Nashville another 6 hours, 364 miles away. He was arrested in a mundane traffic stop in his all-black Hyundai Elantra that displayed a Confederate States of America bumper decoration around the South Carolina license plates while traveling on US Route 74, once known as the "American Indian Highway" (see Figure 6.1). This mundaneness represents the same banality of evil that Arendt (1963a) was trying desperately to convey to readers in her account of the trials of Adolph Eichmann held in 1961 in Jerusalem. After an exhaustive series of taped interviews, "it was as though in those last minutes [Eichmann] was summing up the lesson that this long course in human wickedness had taught us – the lesson of the fearsome, word-and-thought-defying *banality of evil*" (Arendt, 1963a, p. 252). Arendt's searing analysis argued that "despite all of the efforts of the prosecution, everybody could see that this man was not a 'monster,' but it was difficult indeed not to suspect that he was a clown" that Eichmann was vapid of remarkability, pedestrian in his knowledge and conventional in his motivations (Arendt, 1963a, p. 54). It is with that reality that this essay situates Dylann Roof firmly in the coming of age and travel essay narratives. Those narratives position Roof and his activities into the tourism literature, and by doing so, proffer an indicting ideological critique on: this literature, tourism as a concept in society and the myths of sanctity, salvageability and socialisation.

The fraught realities in "Coming of Age" narratives and travel essays

"Coming of Age" narratives rarely present a "coherent narrative", and so are stories "with no settled meaning or ending" (Rogers, 2005, p. 262). Ironically but also fittingly, *The Diary of Anne Frank* and her harrowing experiences during the Holocaust may have been commandeered by readers to mean a host of things, including "triumph, hope, and innocence" (Rogers, 2005, p. 262). Frank as a real person was converted into a character that "grows up" through conflict. It stands now and henceforth as a "Holocaust document", as her story was cut short by the actions of Eichmann and his colleagues in carrying out the "Final Solution". But Frank was not crafting a story for us to repurpose.

She was simply surviving through writing, and as readers we are witnessing this feat of survival. But our embrace of this feat and story forecloses our full understanding and analysis of the Holocaust as a system and process of the Nazi German State. We focus on Frank, not Nazi Germany, and so we are mesmerised by the possibility of hope rather than the futility linked with despair. This is problematic for both children and adult readers of *The Diary.* This is true of most coming of age narratives and memoirs, this failure to see the larger structures concurrently operating in society as well as the direct or indirect manufacturer of the story. These "Coming of Age" stories, whether fiction or non-fiction, are presented as links for people to connect across cultures and time periods. In the course of his trial, Eichmann recounted his days in the YMCA that initiated his pursuit of seeking and joining anything, which ultimately led to his membership in Reich as an adult. In the course of his webpage and travels, Roof recounted his developing understanding of the issues of the day in the world tied to Race. For both Eichmann and Roof, these experiences fell within *the beginning* pattern of these narratives that moves towards *the problem* pattern where the challenge occurs, followed by *the struggle* to overcome that challenge pattern. Roof's journaling and travel itinerary suggested a form of building courage, "testing the waters" and fact-finding. This struggle led to a *big decision*, a pattern rife with tough decisions and obstacles. Roof had settled on a site, drove to it, entered the building, did what had to be done and carried it out after sitting with his victims during their meeting. Unlike Frank, Roof was the star of his own show as he knew he was making a memoir for an audience; an audience that he had also grown frustrated with because of the lack of conviction and action, so maybe his actions could encourage others. Unlike

Frank, Roof was afforded *the ending* pattern that allowed him to become, to be different than he was.

Travel essays serve to some degree the same function as "coming of age" narratives, as they work to describe a particular journey for an author or character and the influence that this journey has had upon them. While the "coming of age" narrative focuses on the development in the midst of a challenge, thereby in the middle of the story, the travel essay emphasises the impact of the travel through some form of recall and processing. The travel essay is the end. In true fashion of this age of social media, and just like many of his media influencing contemporaries, Roof not only pronounced his journey, the comings and goings all along the way, but he also posted his processes of what he had learned. Travel essays are the writings in voyage of personal discovery, and travel and tourism give that voyage style and form. At each site and stop, the author of the travel essay can ramble on about what they see and feel, so that we too can travel with them as they are or have travelled. In different cities and different parts within cities, "there are...unspoken rules of traffic [that] are suspended and you go with a different flow...you can create your own sense of place, give your life a new kind of movement" (Gbadamosi, 1999, p. 185). As another twist of irony, Gbadamosi's (1999) musing on postcolonial Africanness and Blackness of the streets in Brixton in the midst of the Britishness in the greater city London can also serve a reference point for Roof's developing Whiteness and White Nationalist-ness. For Gbadamosi (1999), "Brixton, where [he] lives, is a source for [his] writing about [his] culture" (p. 190), while for Roof, South Carolina was the source of his writings and travels beyond the home. Roof exemplified in his own travel writing the success of what Homi K. Bhabha defined as the primary objective of colonial discourses, to render the colonised as a "population of degenerate types on the basis of racial origin, in order to justify conquest and to establish systems of administration and instruction" (1994, p. 70). However, Roof learned through his readings and observations that Black subjugation was insufficient for a future in the United States; instead, the realisation of extermination was now required as solution. As a comparison, Stein (2009) posited a critique of the travel essays during a hike (*hi-tiyul*) undertaken by Yizak Ben Zvi, who would become Israel's second president: the essay should be read as a travelogue to establish a Zionistic claim to a predominately Arab landscape. Travelogues can and have served nationalistic endeavours while being masked within legitimate ethnic experiences. For Stein, the sensory explorations inherent in the *tiyul* served as "instructional tools and acts of conquest that provided the means for active reclamation of the national homeland through bodily contact with the landscape and cognitive mastery of its contours" (2009, p. 337). And in the course of those travels, the "bad Arab" was a necessary encounter to be articulated as an "essential category". The pursuit of power and dominance over others lulls you into finding the key differences that a system of media, educational and government messaging can exploit. For

Roof, all it took was for him to "type in the words 'black on White crime' into Google, and I have never been the same since that day" (Robles, 2015). Some of these "facts" appeared on a website of the Council of Conservative Citizens that contained "pages upon pages of these brutal Black on White murders". He questioned, "how could our faces, skin, hair, and body structure all be different, but our brains be exactly the same?". For Roof, the skin fetish of Race was not simply for the purpose of gazing and judgement; it was the mark of metaphorical degeneracy of a society and the justification for an enacting of ideology. His travelogue incorporated *the collage effect* that highlighted the transitions from one locale to the next within the context of the author's state of mind, and *the hinge* that situated and shifted the point of view.

Despite the preponderance of fictional and non-fictional, geographically relevant "coming of age" narratives and travel essays, there is a concerning paucity in the presence of these narratives and essays as a genre or subject matter of study in tourism studies literature with most works published before 2000. The solo travel trip of discovery, particularly the road trip, is discussed as an early effort by government to intentionally develop scenic roads and parkways (Morrison, 1969); requiring the assistance of road maps that could be acquired at state information centres as replacements of travel guides (Sandler, 1977); with benignity about the "leisurely pursuit" of back road traveling (Smith & Smith, 1978); indirectly in the historical examination of tramping, the working class youth answer to the aristocratic Grand Tour (Adler, 1985); to a lesser extent, as part of the culture of international long-term budget travellers (Riley, 1988); and within a study of travel to remote sites through the determination of ideal routes to the site and the return home (Taplin & Qiu, 1997). The only mention of and in association with road trips/car travel after 2000 was a book review by Gibson and Connell's *Music and Tourism: On the Road Again* (Butler, 2006). It should be noted that car travel is implied in several articles, particularly the precarity in traveling in the Southern portion of the United States by Black and other travellers of colour (Alderman, 2013, 2018; Alderman et al., 2016; Benjamin et al., 2016; Carter, 2008; Dillette & Benjamin, 2017; Duffy et al., 2019; Hudson et al., 2020; Lee & Scott, 2017; Torabian & Miller, 2017). However, none of these sources explicitly place the discussion exclusively on this form of travel or spend time discussing any of the unique aspects of this form of travel. But more importantly, while virtually all acknowledge or focus on the implications of the racial violence on the travel behaviour of violence receivers, none deal with the violence performers. This oversight in research that many of us have contributed to perpetuates the same error in the representations of *The Diary of Anne Frank* (Rogers, 2005), as mentioned earlier, an exclusive focus on the micro (the victim, survivor or the dead) and not the macro level (the killing machine, the killing system). As a result, we celebrated the empty victory in the removal of Confederate flags and monuments in response to the Charleston deaths as the end rather than the beginning of the facing of truth, reconciliation and restitution that must occur (Webster & Leib, 2016). In our error and our overemphasis, the magnitude of violence is perceived to reside predominantly in the past or is reduced to the actions of present-day isolated individuals or lone actors.

Mapping the travails of Dylann Roof

The site of the killings, the Emanuel African Methodist Episcopal Church, was organised in 1791 and formed as a separate congregation in 1816 and is considered one of the oldest (if not the oldest) Black Church in the US South. Both its constituencies and the building itself have endured a great deal over the centuries. The original Church was once insinuated as harbouring the Denmark Vesey-led meetings on the instigation of slave insurrections and nearly completely burned down in 1822. Rebuilt in 1834, the congregation met in secrecy for fear of death until 1865. The present structure was built in 1891 after enduring an 1886 earthquake (see Figure 6.2; National Park Service, 2018). And it is with this history that some believe the site of the killings was selected. Although others attest that Roof was not thinking of the Church as "his primary target because [one of his friends said that] he was going for the school [UCA University of Charleston]...[but] couldn't [get] into the school because of the security", an explanation that was also given to one his Black friends (Krol, 2015). But under arrest and

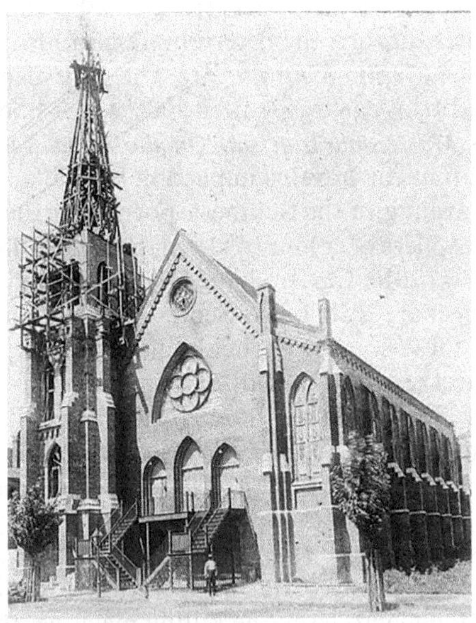

Figure 6.2 Emanuel African Methodist Episcopal Church in 1909. Historical Charleston Foundation/Public Domain.

questioning by the Federal Bureau of Investigations (FBI), Roof spoke on his selection of the Church because of its historical significance, the lack of "real" White Nationalist organisations in Charleston and knew that on that weekday night only Black people would be at a Bible study (Smith et al., 2020). He also researched this information on the internet, using the terms AME churches on sciway.net, a South Carolina "information highway" that is most useful for travel. If he hadn't killed, would this travel activity still be concerning? Yes, but likely unimportant. If he hadn't been a White Nationalist and just posted about his travails, should this travel activity still be concerning? Yes. But why? It is the fact that with just simple substitutions and tweaks or the complete removal of the killing act that levies a scathing of all-things tourism. Roof's travel activities included online trip planning for the uninformed traveller (Smith et al., 2009); user-generated travel posting to share experiences (Wilson et al., 2012); the increasing interest and marketing of Black and African history-based heritage tourism, among Black travellers (Gordon, 2015; Thomas, 2009) and White travellers (Teye et al., 2011); and the selfie and travel (Dinhopl & Gretzel, 2016), especially at sites of atrocity (Cui et al., 2020), showed that Roof was every bit everyday, ordinary and mundane. And in being everyday, with the use of the design work of Tan (2015) in a *Washington Post*'s reporting of his travels, we saw that Dylann Roof moved through the states of South and North Carolina, learning, experiencing, performing and deciding. Through both web captures of past postings from his website as well as the testimony of FBI agent Joseph Hamski in Roof's trial, we have a greater perspective of his road trips and the repeat visits and stops to Mother Emanuel from late 2014 up to the night of the killings (Darlington, 2020).

We are aware by the website posting activity on lastrhodesian.com that he visited several places, visually depicting at least eight other sites. Within the city of Charleston, Mother Emanuel AME sits 115.6 miles away from his home in Columbia, South Carolina. Roof also visited Magnolia Plantation and Gardens 109.1 miles away (see Figure 6.3), a site deemed as one of "America's Most Beautiful Gardens" in *Travel + Leisure Magazine* according to the Planation's website (Magnolia Plantation & Gardens, n.d.), and has been open to the public since 1870. In a society that has maintained and increased the viewing of these sites as destinations for beauty, gardening, romance and matrimony (Buzinde, 2010), it is difficult to situate a completely accurate portrayal of Roof's visitation to this site and understand as anything no more bizarre than the thousands of other individuals who ventured here (see Figure 6.4).

If this was done in the same day and in this order, it took 117.9 miles to travel from Columbia to the McLeod Plantation Historic Site that was also in Charleston. Managed by the Charleston County Parks Department, the site and location took a very different tone in presentation. Whereas Magnolia with its long history of slaving since 1676 before the formation of the United States, McLeod Plantation was established in 1851 during the

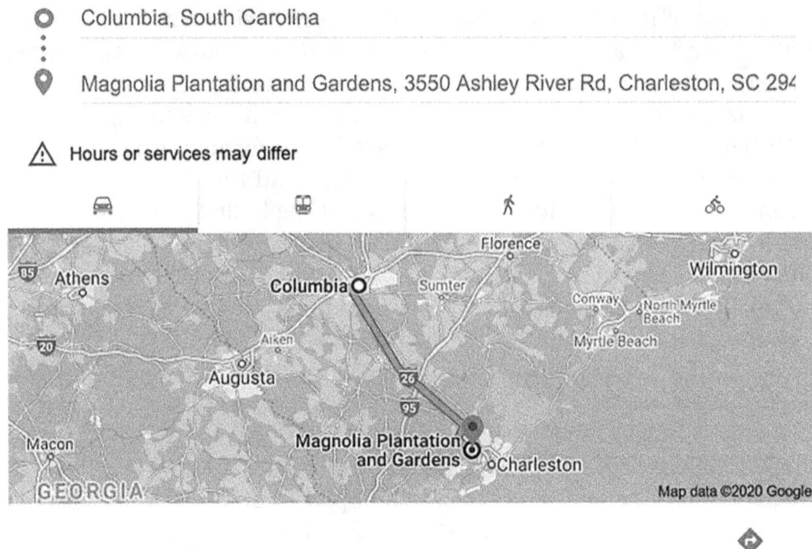

Figure 6.3 Travel distance from Columbia, South Carolina, to Magnolia Plantation and Gardens.

Figure 6.4 One of Dylann Roof's trips to Magnolia Plantation on 23 April 2015. Charleston, South Carolina. Photo Credit: Alamy.

peak of "American" slavery; it was turned not just into "a place for memorialization and a place of conscience, but a place where the transformation of conscience can occur" (Charleston County Parks, n.d.). Admission for a then 21-year-old Dylann Roof was likely US$15, with the option of guided or app-driven, self-guided interpretative tours (see Figure 6.5). But even with a respect for history and the presence of docent providing rich histories to visitors (Modlin, 2008), reviews of visitations by White travellers remarked on the "heavy bias" and "embellishing" on the wrongs of slavery and leaving with feelings of being attacked at McLeod site. Some Black re-enactors at other sites have noticed that at least "once a week or so, visitors walk away after realizing they're talking with a man playing a slave. Sometimes they sigh or say, 'Not this again'" (Knowles, 2019). Roof visited this site once on 23 April 2015 and then again on 9 May 2015, when he also visited Mother Emanuel AME.

This move towards realism and historical representation has led to some plantation sites, like Boone Hall Plantation and Gardens (2020), to become highly commercialised sites. While it is recognised as the #1 Plantation in the Charleston area by *USA Today*, it is still a fully functional plantation, albeit without slave labour (see Figure 6.6). Founded in 1681, it caters to both plantation weddings enthusiasts and those seeking educational opportunities. The site, 120.4 miles away from Columbia, likely featured docents, re-enactors and wax figurines of enslaved peoples, some of whom

Figure 6.5 One of Dylann Roof's trips to McLeod Plantation Historic Site on 23 April 2015. Photo Credit: Alamy.

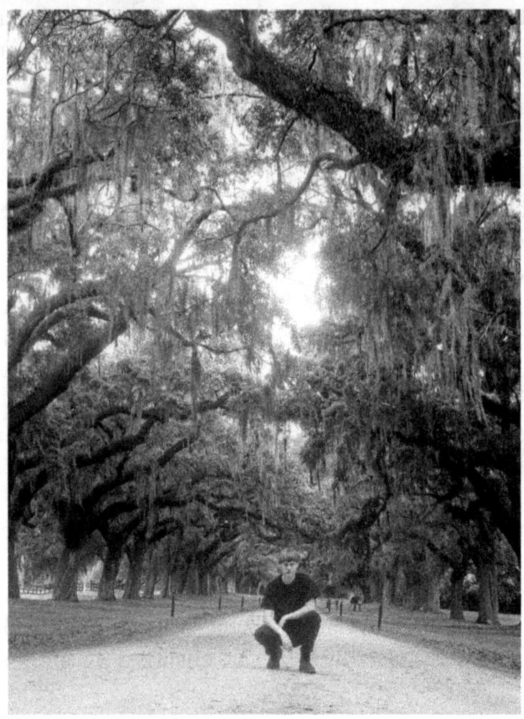

Figure 6.6 Dylann Roof on 17 June 2015 at Boone Hall Plantation. Mt. Pleasant, South Carolina. Photo Credit: Alamy.

Roof posed in a series of photos with (see Figure 6.7). Among many photos, one showed Roof missing the timed snap of the picture, grinning with eyes closed behind two figurines. Roof visited this site on 22 December 2014, 13 April 2015 and again on 25 April 2015, when he also visited Daniel Island and Mother Emmanuel in Charleston where he travelled to in December and earlier in April.

Some 124.9 miles away from Columbia is the 400-year-old Angel Oak site on John's Island, which was likely visited by Roof on 27 February 2015 (see Figure 6.8). He also visited Sullivan's Island and Fort Moultrie on the same day, roughly 122.1 miles from Columbia (see Figure 6.9). Sullivan's Island under British Colonial rule was the principal location for the disembarking of nearly 40% of all enslaved persons in the Colonies; nearly 360,000 persons are accounted for in the ship logs and market purchases. He returned to Sullivan's Island on 16 March 2015, where he marked in sand "1488". This is White Nationalist code for the 14 words "We must secure the existence of our people and a future for White children" and "88" for Heil Hitler as the letter "H" is the eighth letter of the alphabet. As Michael postulated, "the 14 Words [are] a litmus test for activists in the movement. If they cannot

Figure 6.7 Dylann Roof on 3 August 2014 at a slave cabin museum exhibit at the Boone Hall Plantation. Mt. Pleasant, South Carolina. Photo Credit: Alamy.

Figure 6.8 Dylann Roof on 16 March 2015 at Angel Oak. John's Island, South Carolina. Photo Credit: Alamy.

Figure 6.9 Dylann Roof on 16 March 2015 at Sullivan's Island. Photo Credit: Alamy.

endorse the [David Lane's, the creator of the] 14 words without equivocation, then their sincerity and reliability are called into question" (2009, p. 51). The 14 words are not just an actuation of realised White power; it is a necessary call to action to prevent the destruction of a Race. White power is an achievement in striving for White Nationalism, while also serving as an underpinning of White Supremacy. Later, on the 27 February 2015, and after picking up a bite to eat at Chick-fil-A in Mount Pleasant, South Carolina, he returned back to Mother Emanuel and then headed home to Columbia. He drove to Mother Emanuel only three days prior, on 24 February 2015. Among his other stops, Roof was 103.6 miles away in Greenville, South Carolina at the Museum and Library of Confederate History and the Sterling Statue on 10 May 2015 (see Figures 6.10 and 6.11). On 16 May 2015, Roof also went to Kensington Plantation 24 miles away in Eastover, South Carolina. He routed the distance to Charleston and stopped near Mother Emanuel AME, according to GPS logs. It has not been noted on what dates Roof travelled to nearby Elmwood Memorial Gardens, also known as the Confederates Soldiers Cemetery and Finley Park. Regardless, over the span of several months Dylann Roof visited and frequented a range of Capitalist heritage-based tourism, African American Historical and White Nationalist sites.

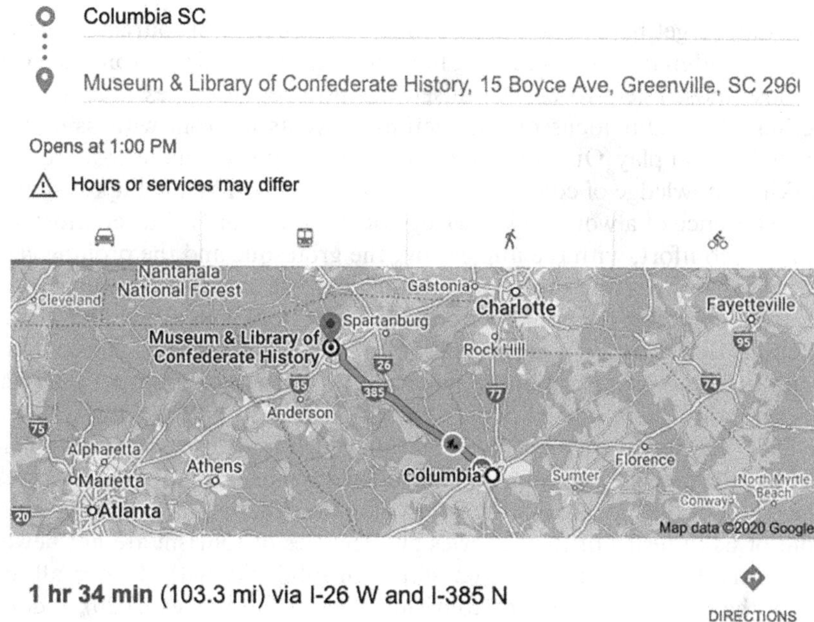

Figure 6.10 Travel distance from Columbia, Museum and Library of Confederate History in Greenville, South Carolina.

Figure 6.11 Dylann Roof on 24 April 2015 at the Museum and Library of Confederate History. Greenville, South Carolina. Photo Credit: Alamy.

The banality of white nationalist tourism

How did we get here? The ways in which we conceive of tourism, even in criticism, often fails to take a long look into injustice and not from the view of those most likely to be victimised but instead the forces that renders victims of us. Our focus on the victims prevents us from witnessing the larger forces at play. Our conception of decolonisation is in the absence of a working knowledge of colonialism. Our conception of White Supremacy is in the absence of a working knowledge of those 14 words. Our comfort, or really discomfort, with the unpleasant, the grotesque and the profane gets in the way of inverting the gaze of trauma experiences away from trauma-inducing outcomes. Arendt cautioned us to not look away, but also to understand that "justice does not permit anything of the sort; it demands seclusion…and it prescribes the most careful abstention from all the nice pleasures of putting oneself in the limelight" (Arendt, 1963b). Also, why are we here? As in why are we at this point that such immense energy and tacit focus is devoted to calls for representation, inclusion and social justice? The issues of the day are not the issues of many other yesterdays; what is in front of us in terms of the histories and legacies of tourism are not newly known truths. While this chapter stands in solidarity with the overall focus of the book on socialising tourism (Higgins-Desbiolles, 2020), it does present a broadening and a challenge. Can we truly engage in socialising tourism without an organised program on truth, reconciliation and restitution that interrogates the reality that tourism does not have safeguards for less harmful, more unjust use? What is truth when a society manufactures "systematic mendacity" (Arendt, 1963b) that makes State and State sanctioned actions so easy? In proposing a more organised programme centred on truth in tourism, an emphasis here on establishing studies of histories situates the context of racial violence and ultimately the cleansing of populations to create more narratives that are essential to creating more meaningful societies.

Beyond the actions of the State in the repression of society, the State also sanctions the actions of vigilantes and organised militias to commit neoliberal injustices. To control the expenditures of budgets and to look the other way, Roof and other actors have served a function to strike fear, but to also do the work of elimination by animating leisure spaces (Johns, 2017; Mowatt, 2019) and by using tourism and travel to engage their targets and enact their cause. In an early period of the internet, Benjamin Smith in 1999 carried out a touring shooting spree between two states, Illinois and Indiana, but across five cities that lead to the death of Northwestern University's first Black Basketball coach and a newly admitted Korean Doctoral Student at Indiana University (see Figure 6.12). However, instead of situating these acts as those of individuals who are simply "bad apples", I posit within this chapter that the racial project of White Nationalism has grown and flourished with the internet and has, in fact, benefitted from the anonymity and distance that these digital cultures and landscapes afford them and their intentions, at our peril. The lasting impact of the 14 Words can be seen on 27 October 2008, when Paul Schlesselman, age 18, of Arkansas, and Daniel Cowart, age 20, of Tennessee, plotted to assassinate the 44th President of the United States, and would have culminated in the killing of 88 Black people across multiple state lines, including 14 beheadings. Two White Nationalist cosplayers, Jerad Miller (age 31, as the Joker) and Amanda Miller (age 22, as Harley Quinn) fatally shot two cops and a bystander in Las Vegas, Nevada, after travelling across country from Indiana. The availability of content on the internet and its influences was shown on 22 March 2017, as James Harris Jackson, age 26, travelled from Baltimore to New York with the express intent of killing as many Black men as possible to sway White women from engaging in interracial relationships. The public stabbing of 66-year-old Timothy Caughman with a two-foot sword was meant to be only a test run. The 17-year-old Kyle Rittenhouse was driven by his mom from Antioch, Illinois, to protect property and exact vengeance against protestors who were protesting against the killing of Jacob Blake in Kenosha, Wisconsin, 21 miles away; Rittenhouse killed two protestors and injured one in the late evening of 23 August 2020 (Miah, 2020). The online activity of White Nationalists, in particular through their cloaked websites, offers an ability to hide political agendas, conceal authorship, obscure the identification of knowledge production, generate false or inaccurate content and hinder punitive accountability for eventual actions taken by these "virtual tourists" turned ethnic cleansing road trippers. And there is enough

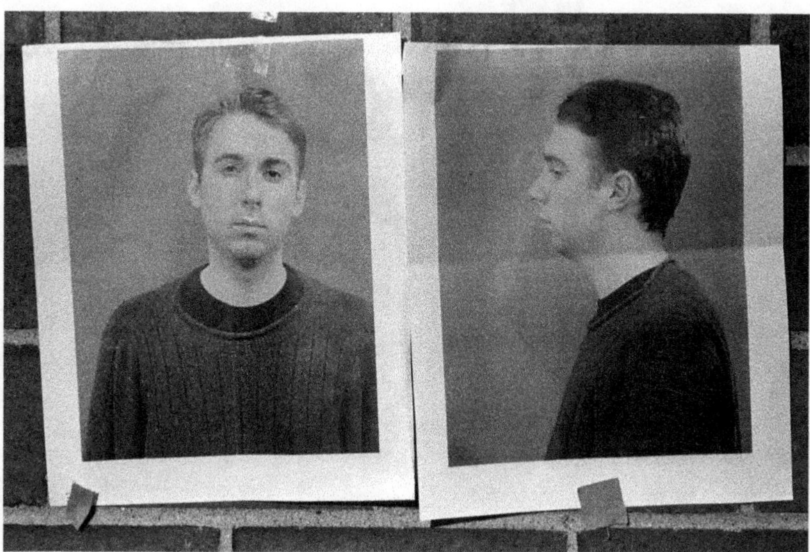

Figure 6.12 Benjamin Nathaniel Smith photo posted on the wall of a Chicago Police Station after killing Ricky Byrdsong and shooting six Orthodox Jews on 2 July; 4 July 1999. Photo Credit: Alamy.

evidence that more of these travellers are to come, as a veteran White nationalist Harold Covington noted:

> They've been given a vision of a time in some imagined but possibly not too far distant future when all of a sudden...they will see a young white man like Dylann Roof standing in front of them with no steroid-pumped policemen in blue to protect their liberal candy asses from the consequences of years of their own behavior...They will see in that young White man's eyes...That he is now beyond deception or bullying or browbeating or Twitter-shaming or intimidation, that he knows them for what they are. And they will look down and see that he has something in his hand (cited in Thielman, 2015, n.p.).

Conclusion

Dylann Roof travelled in his mind as much as he travelled via car. He made note, questioned by researching, "deeper and found what was happening in Europe...I saw that the same things were happening in England and France, and in all the other western European countries" (Smith et al., 2020). He watched movies like *Made in Britain 1982* and used images on his website from another movie *Romper Stomper* about skinheads but set in the United Kingdom and Australia. He ordered patches displaying the flags of Apartheid South Africa and Rhodesia (present-day Zimbabwe) and sewed them on his jacket that he proudly wore in many of his selfies (see Figure 6.13).

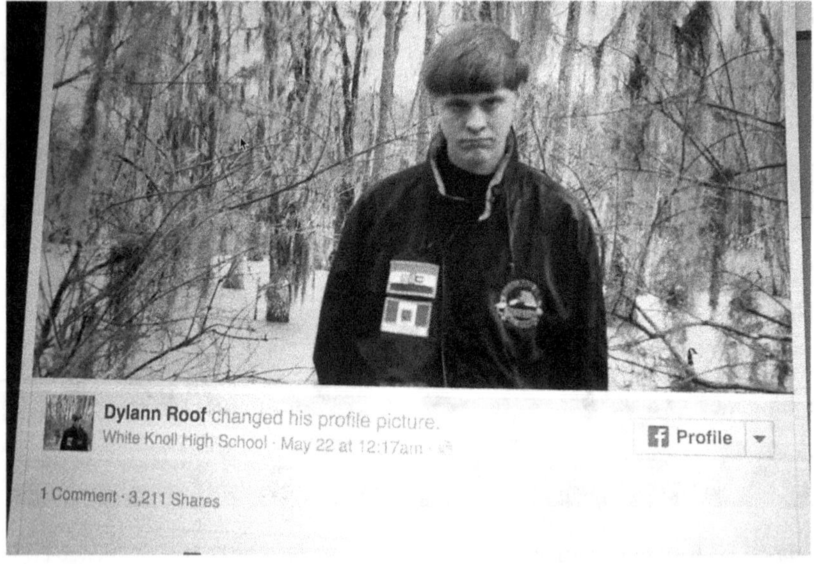

Figure 6.13 Dylann Roof's FaceBook post on 22 May 2015 of a picture at the Magnolia Plantation. Photo Credit: Alamy.

Figure 6.14 Dylann Roof in a restroom prominently displaying Apartheid South Africa and Rhodesia flags, taking a selfie on 18 March 2015. Photo Credit: Alamy.

Arendt (1963b) observed and critiqued all who watched the Eichmann trial, because we do not want "to admit that an average, 'normal' person, neither feeble-minded nor indoctrinated nor cynical, could be perfectly incapable of telling right from wrong" (Arendt, 1963b). A person could be vapid of anything, compassion or hatred. After engaging in a road trip, stopping for gas and the restroom selfies (see Figure 6.14) over the span of seven months, Roof would empty 77 rounds into nine people and remain in the building for 45 minutes to get some rest before driving on to Nashville; this was just a person doing what everyone does, travel. Even embarrassed by his manifesto by closing it with "please forgive any typos…I didn't have time to check it", before he began his 113.7-mile drive.

References

Adler, J. (1985). Youth on the road: Reflections on the history of tramping. *Annals of Tourism Research, 12*(3), 335–54. https://doi.org/10.1016/0160-7383(85)90003-9

Alderman, D. (2013). Introduction to the special issue: African Americans and tourism. *Tourism Geographies: An International Journal of Tourism Space, Place and Environment, 15*(3), 375–9. https://doi.org/10.1080/14616688.2012.762544

Alderman, D. H. (2018). The racialized and violent biopolitics of mobility in the USA: An agenda for tourism geographies. *Tourism Geographies, 20*(4), 717–20. https://doi.org/10.1080/14616688.2018.1477168

Alderman, D. H., Butler, D. L., & Hanna, S. P. (2016). Memory, slavery, and plantation museums: The river road project. *Journal of Heritage Tourism, 11*(3), 209–18.

Arendt, H. (1963a). *Eichmann in Jerusalem: A report on the banality of evil.* New York: Viking Press.

Arendt, H. (1963b, February 9). Eichmann in Jerusalem. *The New Yorker.* Retrieved 6 November 2020, from https://www.newyorker.com/magazine/1963/02/16/eichmann-in-jerusalem-i

Benjamin, S., Kline, C., Alderman, D., & Hoggard, W. (2016). Heritage site visitation and attitudes toward African American heritage preservation: An investigation of North Carolina residents. *Journal of Travel Research, 55*(7), 919–33. https://doi.org/10.1177/0047287515605931

Bhabha, H. K. (1994). *The location of culture.* London: Routledge.

Boone Hall Plantation & Gardens (2020). *A must see stop on any trip to Charleston, S.C.*. Retrieved 7 November 2020, from https://www.boonehallplantation.com/

Butler, R. W. (2006). Music and tourism: On the road again. *Tourism and Hospitality Research, 6*(2), 188–91. https://doi.org/10.1057/palgrave.thr.6040056

Buzinde, C. N. (2010). Discursive constructions of the plantation past within a travel guidebook. *Journal of Heritage Tourism, 5*(3), 219–35. https://doi.org/10.1080/1743873X.2010.508525

Carter, P. L. (2008). Coloured places and pigmented holidays: Racialized leisure travel. *Tourism Geographies, 10*(3), 265–84.

Charleston County Parks (n.d.). *McLeod Plantation Historic Site.* Park & Program Services. https://www.ccprc.com/1447/McLeod-Plantation-Historic-Site

Cobb, J. (2017, February 6). Inside the trial of Dylann Roof. *The New Yorker.* Retrieved 15 February 2017, from http://www.newyorker.com/magazine/2017/02/06/inside-the-trial-of-dylann-roof/amp

Cui, R., Cheng, M., Xin, S., Hua, C., & Yao, Y. (2020). International tourists' dark tourism experiences in China: The case of the memorial of the victims of the Nanjing Massacre. *Current Issues in Tourism, 23*(12), 1493–511. https://doi.org/10.1080/13683500.2019.1707172

Darlington, A. (2020, September 4). Prosecution's timeline of Dylann Roof's movements. *The Post and Courier.* Retrieved 6 November 6 2020, from https://www.postandcourier.com/prosecution-s-timeline-of-dylann-roofsmovements/article_a9a6fb38-c190-11e6-89ef-d7a3ed7ab88b.html

Dillette, A., & Benjamin, S. (2017). Traveling while Black: Storytelling through Twitter. *Critical Tourism Studies Proceedings, 1*, 110.

Dinhopl, A., & Gretzel, U. (2016). Selfie-taking as touristic looking. *Annals of Tourism Research, 57*, 126–39. https://doi.org/10.1016/j.annals.2015.12.015

Duffy, L. N., Pinckney, H. P., Benjamin, S., & Mowatt, R. (2019). A critical discourse analysis of racial violence in South Carolina, U.S.A.: implications for traveling while Black. *Current Issues in Tourism, 22*(19), 2430–46. https://doi.org/10.1080/13683500.2018.1494143

Gbadamosi, G. (1999). The road to Brixton Market: A post-colonial travelogue. In S. H. Clark (Ed.), *Travel writing and empire: Postcolonial theory in transit* (pp. 185–94). London: Zed Books Ltd.

Gordon, T. S. (2015). 'Take Amtrak to Black history': Marketing heritage tourism to African Americans in the 1970s. *Journal of Tourism History, 7*(1–2), 54–74. https://doi.org/10.1080/1755182X.2015.1047804

Higgins-Desbiolles, F. (2020). Socialising tourism for social and ecological justice after COVID-19. *Tourism Geographies, 22*(3), 610–23. https://doi.org/10.1080/14616688.2020.1757748

Hudson, S., So, K. K. F. S., Meng, F., Cárdenas, D., & Li, J. (2020). Racial discrimination in tourism: The case of African-American travellers in South Carolina. *Current Issues in Tourism, 23*(4), 438–51. https://doi.org/10.1080/13683500.2018.1516743

Johns, A. (2017). Flagging white nationalism 'after Cronulla': From the beach to the net. *Journal of Intercultural Studies, 38*(3), 349–64. https://doi.org/10.1080/07256868.2017.1314259

Knowles, H. (2019, September 8). As plantations talk more honestly about slavery, some visitors are pushing back. *The Washington Post*. Retrieved 6 November 2020, from https://www.washingtonpost.com/history/2019/09/08/plantations-are-talking-more-about-slavery-grappling-with-visitors-who-talk-back/

Krol, C. (2015, June 20). Dylann Roof's friend: Charleston church "wasn't primary target". *The Telegraph*. Retrieved 10 February 2017, from https://www.telegraph.co.uk/news/worldnews/northamerica/usa/11688181/Dylann-Roofs-friend-Charleston-church-wasnt-primary-target.html

Lee, K. J., & Scott, D. (2017). Racial discrimination and African Americans' travel behavior. The utility of habitus and vignette technique. *Journal of Travel Research, 56*(3), 381–92. https://doi.org/10.1177/0047287516643184

Magnolia Plantation & Gardens (n.d.). *Charleston's most visited plantation*. Retrieved 9 November 2020, from https://www.magnoliaplantation.com/

Miah, M. (2020). Shooting of Jacob Blake intensifies racial divide [online]. *Green Left Weekly, 1279*, 14–15. Retrieved 6 November 2020, from https://search.informit.com.au/documentSummary;dn=406608717422892;res=IELHSS

Michael, G. (2009). David Lane and the Fourteen Words. *Totalitarian Movements and Political Religions, 10*(1), 43–61.

Modlin, E. A. (2008). Tales told on the tour: Mythic representations of slavery by docents at North Carolina Plantation Museums. *Southeastern Geographer, 48*(3), 265–87.

Morrison, C. C. (1969). A proposed program for scenic roads and parkways. U.S. Department of Commerce. *Journal of Leisure Research, 1*(1), 91–94. https://doi.org/10.1080/00222216.1969.11969715

Mowatt, R. A. (2019). A people's history of leisure studies: Where the white nationalists are. *Leisure Studies*. https://doi.org/10.1080/02614367.2019.1624809

National Park Service (2018, February 22). *Emanuel A.M.E. Church*. National Park Service, The U.S. Department of the Interior. Retrieved 3 November 2020, from https://www.nps.gov/places/emanuel-a-m-e-church.htm

Riley, P. J. (1988). Road culture of international long-term budget travelers. *Annals of Tourism Research, 15*(3), 313–28. https://doi.org/10.1016/0160-7383(88)90025-4

Robles, F. (2015, June 20). Dylann Roof photos and a manifesto are posted on website. *The New York Times*. https://www.nytimes.com/2015/06/21/us/dylann-storm-roof-photos-website-charleston-church-shooting.html

Rogers, T. (2005). Understanding in the absence of meaning: Coming of age narratives of the Holocaust. *The New Advocate, 15*(4), 259–66.

Sandler, M. (1977). Road maps: Handle with care. *Journal of Travel Research, 16*(1), 23–25. https://doi.org/10.1177/004728757701600107

Smith, C. J., & Smith, C. A. (1978). Hodography: A plea for the back roads. *Leisure Sciences, 1*(4), 411–26. https://doi.org/10.1080/01490407809512898

Smith, G., Hawes, J. B., & Darlington, A. (2020, September 14). Dylann Roof says he chose Charleston, Emanuel AME for massacre because they were historic, meaningful. *The Post and Courier.* Retrieved 3 November 2020, from https://www.postandcourier.com/church_shooting/dylann-roof-says-he-chose-charleston-emanuel-ame-for-massacre-because-they-were-historic-meaningful/article_6fab532c-be05-11e6-ab05-575a173993ee.html

Smith, W. W., Pan, B., Li, X., & Zhang, G. (2009). Conceptualizing the impact of geographical ignorance on online trip planning. *Tourism Geographies, 11*(3), 350–68. https://doi.org/10.1080/14616680903032775

Stein, R. L. (2009). Travelling Zion: Hiking and settler-nationalism in pre-1948 Palestine. *Interventions, 11*(3), 334–51. https://doi.org/10.1080/13698010903255569

Tan, S. (2015, July 1). Map of Dylann Roof's travels. *The Washington Post.* Retrieved 12 February 2017, from https://www.washingtonpost.com/graphics/national/dylann-roof-travels/

Taplin, J. H. E., & Qiu, M. (1997). Car trip attraction and route choice in Australia. *Annals of Tourism Research, 24*(3), 624–37. https://doi.org/10.1016/S0160-7383(97)00021-2

Teye, V., Turk, E., & Sönmez, S. (2011). Heritage tourism in Africa: Residents' perceptions of African American and White tourists. *Tourism Analysis, 16*(2), 169–85. https://doi.org/10.3727/108354211X13014081270404

Thielman, S. (2015, June 28). White supremacist calls Charleston "a preview of coming attractions". *The Guardian.* Retrieved 8 February 2017, from https://www.theguardian.com/us-news/2015/jun/28/harold-covington-northwest-front-dylann-roof-manifesto-charleston-shooting

Thomas, L. (2009). "Roots Run Deep Here": The construction of Black New Orleans in post-Katrina tourism narratives. *American Quarterly, 61*(3), 749–68.

Torabian, P., & Miller, M. C. (2017). Freedom of movement for all? Unpacking racialized travel experiences. *Current Issues in Tourism, 20*(9), 931–45. https://doi.org/10.1080/13683500.2016.1273882

Webster, G., & Leib, J. (2016). Religion, murder, and the Confederate Battle Flag in South Carolina. *Southeastern Geographer, 56*(1), 29–37.

Wilson, A., Murphy, H., & Fierro, J. C. (2012). Hospitality and travel: The nature and implications of user-generated content. *Cornell Hospitality Quarterly, 53*(3), 220–28. https://doi.org/10.1177/1938965512449317

7 Dismantling the ivory tower

A narrative ethnography between two critical scholars

Stefanie Benjamin and Alana Dillette

Introduction

It was the morning of 25 June 2020. It had already been a tough month. In fact, this day marked exactly one month since the brutal killing of George Floyd and subsequent global social unrest in response to the continued racial violence against Black bodies in the United States, tourism being no exception (Ferguson, 2020; Spinks, 2020; Worland, 2020). On this morning, we awoke to that familiar pang of anxiety you feel when you notice the subject line "decision on manuscript ID 4567" in your inbox. In this case, the manuscript was one we had submitted months earlier in April (before George Floyd's death) which qualitatively explored Black travel combining critical race and social movement theories. As expected, reviewer comments ranged in criticality and suggestions to improve the manuscript. However, one comment stood out; it read:

> I would suggest submitting this manuscript to a more specialized journal about inequalities but not in a top tourism journal because the link to tourism issues is not very strong. The manuscript has some potential but not for a top-tier tourism journal, but maybe for a more cultural studies/multidisciplinary outlet.

Naturally, reading critical feedback about your work certainly stings. However, this comment felt like a personal attack on critical tourism scholarship; more specifically, scholarship that dares to highlight the inequitable racial underbelly of the travel industry. While it has already been proven that the travel industry and academic landscapes are historically whitewashed and grounded in dominant hegemonic principles of positivist inquiry (Dillette et al., 2019), the need for critical tourism scholarship and pedagogy abounds. As tourism educators and scholars, we have the opportunity and responsibility to challenge these norms through our pedagogical design, our research practices and our relationships with industry collaborators.

DOI: 10.4324/9781003164616-7

As Directors of Tourism RESET – an organisation of academics and practitioners dedicated to the study of issues related to race, ethnicity and social equity in tourism (RESET) – our initiative's goal is focusing on amplifying marginalised voices and advocating for critical, actionable change in the industry and academia. We are working towards dismantling power structures steeped in White supremacy, creating socially equitable touristic landscapes and disrupting the normativity and pervasiveness of Whiteness through education, scholarship and industry collaborations. Challenging these normative power structures is paramount to dismantling the cycle of inequality that permeates academia and society (Arday, 2018), and is even more relevant as we consider how privilege frames dominant narratives traditionally seen and heard within academic and touristic institutions.

The overarching purpose of this chapter is to identify how the structures of colonialism and this cycle of normativity frame our experiences as womxn critical tourism scholars.[1] We critique the underlying hegemonic nature of the tourism industry and higher education to help answer these questions: As scholars, how have we been socialised to exist and perform in academic spaces? How has the COVID-19 pandemic exposed capitalistic structures that permeate academia? In what ways can initiatives like Tourism RESET serve as a blueprint to begin the decolonisation process of curriculum development and research between industry and academia? Informed by narrative ethnography, this chapter offers space for dialogue and reflexivity, evoking emotions, feelings and using the "self" to better digest and critically analyse the world in which we exist (Ellis, 2004). As Cannella and Lincoln (2017, p. 86) posited: "we must struggle to 'join with' and 'learn from' rather than 'speak for' or 'intervene into'". Consequently, we critically analyse our own perpetuation of a hegemonic ideology of our performance within higher education and the tourism industry. Hopefully, by critiquing our work and the landscape that we navigate, we can offer space for deeper reflection and dialogue, creating a framework for equitable, actionable change across tourism and academic landscapes.

The colonisation of higher education

Tourism is one of the largest and most pervasive industries across the globe. Every continent and almost every country partake in some aspect of the tourism industry. Because of the immense amount of money and power involved in this business, studies of tourism have grown exponentially in the last few decades. Reflecting this economic bias in tourism, much of the past research, especially before the mid-to-late 1990s and early 2000s, was conducted in quantitative terms and focused on increasing the bottom line of tourism-related businesses (Walle, 1997). The study of tourism was not unlike other disciplines, as other social sciences went through such a "quantitative revolution" that caused quantitative methods to become dominant in the mid-1900s (Kwan & Schwanen, 2009). Scholars argue that this "quantitative revolution" was influenced by the rise of American economic power, through the institutionalization of neoliberalism, which emphasizes the "quantification of both knowledge production and learning" (Caton, 2015, p. 19). However, right before the turn of the century, critical voices that spoke against the positivistic nature of such quantitative studies arose, arguing for the inclusion of critical perspectives and qualitative methodologies (Riley & Love, 2000; Walle, 1997; Xiao & Smith, 2006). These critical voices, which sought to uncover power relations and inequality that lay hidden in economic statistics, echo the thought that John Tribe laid out in 1997: "Tourism clearly encompasses more than just that which is measurable in monetary terms" (p. 640).

Specifically, scholars influenced by such work have pushed back against positivist methods to take a "critical stance towards taken-for-granted knowledge and understandings that knowledge is historically and culturally specific, that knowledge is sustained by social practices and that knowledge and social action are interlinked" (Tribe, 2008, p. 246). Recognising the influence of neoliberalism in the academy, recent scholars in tourism studies have sought "intellectual enrichment, social justice, and social equity" through their research (Pritchard & Morgan, 2007, p. 21), identifying with Tribe's (2008) description of critical theories that recognise the social production of knowledge. Tracing a lineage back to Jürgen Habermas (1978) and the Frankfurt School in sociology, recent tourism scholars have built on a legacy of interdisciplinary work to introduce critical epistemologies into the field of tourism studies, especially through the recognition of power and knowledge production (Higgins-Desbiolles, 2006). As Tribe writes, "power is a fundamental issue to be researched and a critical approach to tourism would seek to expose whose interests are served and how power operates in particular formations of tourism as well as in the process of research" (2008, p. 246). By understanding the construction and maintenance of power through the tourism industry and the positivist studies of traditional tourism studies, critical scholars are able to explore the implications of privileging such knowledge as well as the oppression of alternative knowledges that may run counter to hegemonic narratives needed to reproduce social and economic conditions that sustain the tourism industry.

The increasing tide of research conducted through such critical epistemologies that question power relations and social constructions in tourism has been dubbed "the critical turn" by several scholars, reflecting the rising prevalence of industry critiques (Ateljevic et al., 2007; Bianchi, 2009). This "critical turn" and its associated scholars "seek to stimulate their audience to transform society and thereby to liberate themselves and others" (Bramwell & Lane, 2014, p. 2).[2] In order to seek this stimulation and transformation, tourism researchers have to recognise tourism as more than an industry, focusing instead on tourism as a social force that has positive impacts, such as "improving individual well-being, fostering cross-cultural understanding, facilitating learning, contributing to cultural protection, supplementing development, fostering environmental protection, promoting peace and fomenting global consciousness, which contributes to the formation of global society" (Higgins-Desbiolles, 2006, pp. 1196–7).

Narrative ethnography

Postmodernists posit that the logical assumptions and scientific methods guide a particular reasoning of the elite's consensual view of reality: the grand narrative (Lyotard, 1984) of Western, mostly White, mostly male, science. As Goodall (2000, p. 13) asserts, "what counts as truth depends on where you are standing in the first place, and what you want to do with it". Furthermore, we write our ethnographic stories for a purpose – and that we locate (from the languages we collect through fieldwork) and invent (out of our professional training and individual sensitives) a "language of contextual meanings for describing, analyzing, and storying a culture as we go along" (Goodall, 2000, p. 13). Thus begins a tradition of narrative analysis centred on how stories reveal the relational selves as orators (Gubrium & Holstein, 2008) and a methodology to identify an in-depth view of a lifestyle of a particular group – using narratives to understand individual experiences through time, especially in relation to significant life transitions.

The method of narrative ethnography aims to closely scrutinise social situations, their actors and actions in relation to narratives, including intensive observation of the field of study (Gubrium & Holstein, 2008). However, more than *just* storytelling, this method provides analytical platforms, tools and sensibilities for capturing the "rich and variegated contours of everyday narrative practice" (Gubrium & Holstein, 2008, p. 251). Most famously, Ellis and Bochner (1996) and Ellis and Flaherty (1992) refer to "narrative ethnography" as their attempt to convey the reflexive, representational engagements of field encounters. Goodall's (2000) book *Writing the New Ethnography* is an exemplary rendition of this form of narrative ethnography. We pull directly from Goodall (2000) informing the flow and design of this chapter – focusing on an *evocative, creative* and *critical* approach to storytelling – especially within the critique of academic culture and dissemination of knowledge:

> There are cultural signs of hope, however. For me, one direct sign of potential change in academic culture is the willingness of some brave academics to turn the critical lens back on our institutions. To examine our cultural practices just as we would any other tribe's. To write about the emotional ups and downs that accompany this otherwise "privileged" way of life. To write, openly, about the cultural politics of publication, tenure, and promotion decisions ... to discuss the role of gender,

> of race, of class, of sexual orientation, in all that we do. To question what we have for so long simply taken for granted.
>
> (p. 29)

Goodall continues to critique the ways in which our academy has been conditioned to present our data and write in a "masculine way", building on "facts" to be used to generalise models, theories or explanations. Consequently, academic research is not seen as a "reactive" expression, but rather for the "common quest for truth" forcing scholarly journals to report the results of principled inquiry within a "tradition of knowledge building that is aimed at representing truth" (Goodall, 2000, p. 59). However, what if truth is gendered in a feminine way or in a non-gendered/cross-gendered way?

As ethnographers, we turn our gaze back to ourselves and begin experimenting more openly with new ways of forming questions, conducting inquiry and writing. Goodall (2000) challenges us as scholars to slow down, really pay attention to what is going on and understand the implications of silence. Thus, Goodall (2000, p. 116) posits three assumptions:

1 We symbolically act in, and on, the world through forms of communication.
2 These symbolic actions are representations of our interpretations of the social world.
3 Patterns of symbolic actions can be organised, understood and represented as cultural performances of everyday life.

These patterns of performance may be coded through routines, rituals and rites of passage. Through this chapter, we tell our stories of our interpretation of academic culture through the process of discovery – where we become part and parcel of the story itself. We are searching for personal and subjective experience in fieldwork – properly seeking reflexivity. In line with Hertz, we imply "a shift in our understanding of data and its collection – something that is accomplished through detachment, internal dialogue, and constant (and intensive) scrutiny of 'what I know' and 'how I know it'" (1997, pp. 7–8).

Reflections of two early career academics

As two assistant professors on the tenure track at institutions in the United States, we share a brief positionality statement of who we are and how we view academia. Within this section, we dive deeper into the questions of how, as scholars, we've been socialised to exist and perform in academic spaces through the major theme: "The tower of the academy". This is inclusive of two subthemes – "Performing as an academic: Publish or Perish" and "The colonisation of tourism knowledge". We end with our second major question, diving deeper into how the COVID-19 pandemic exposed capitalistic structures that permeate academia and the tourism industry through the major theme of "A tale of two pandemics".

The researchers

Stefanie. I identify as a White cisgender, non-disabled womxn growing up in a middle-class American family. I never imagined that I would become a professor – or continue a career within academic studies. However, it was after my time in New York City as an event planner that influenced me to pursue a career with more meaning and substance. I graduated with a PhD in foundational inquiry of education and was taught to critically analyse higher education. I remember my dissertation chair telling me "in order to change the game you have to know how to play the game" – in reference to the politics and power of academia. After graduating, I accepted a position as a tenure-track assistant professor at a Research I institution[3] in the southeastern U.S. I thought that I could be the agent to "change the game" of academia to be more accepting and welcoming of critique and different ways of thinking. However, I was disappointed by the familiar paradigm around "fear of change" that caused me to doubt myself as a scholar and question my position within the system which I was trying so hard to resist. I became frustrated, disappointed and depressed with the institution of higher education and questioned how much longer I could continue to perpetuate my acceptance of this lifestyle that romanticised intense productivity, offered little validation and caused continuous health deterioration.

Alana. I identify as an Afro-Caribbean, mixed race womxn. More specifically, I am Bahamian-Canadian, Black and White, but identify as Black both personally and by society. Currently, I am a pre-tenured assistant professor of hospitality and tourism at a University in Southern California. My journey towards professorship began serendipitously at the beginning of my Master's program when I first became acquainted with tourism scholarship. I became fascinated with the opportunity to critically analyse and evaluate the structures of tourism. Growing up in The Bahamas, tourism had been a part of my vocabulary since childhood. This led me to my first research project critiquing the (un)sustainable nature of tourism development in my home country's archipelago of islands. Following this, I embarked upon my dissertation with the goal of completing an entirely qualitative project. However, I was pushed to use mixed methods as an alternative so that the work would be viewed as more "robust". As a student trying to successfully finish a PhD, I submitted to this assertion of power. In hindsight, I realise the flaw of this position and its roots in the normative power structures of academia and positivistic viewpoints. Now that I am on the "other side" as a professor, I strive to constantly consider tourism education and scholarship through a critical lens, equally inclusive of both qualitative and quantitative methodologies. My biggest fear is that I may unknowingly, or knowingly, push my own colonised ideologies back onto myself or others.

The tower of the academy

Performing as an academic: publish or perish

Within Stefanie's first month of being on campus as an Assistant Professor, she was visited by Vanessa, a Full Professor in her department. Already overwhelmed and stressed out by her academic expectations, this professor felt the need to visit Stefanie in her office and share with her something that she already knew, something that all academics have drilled into their psyches:

VANESSA: I want to extend our excitement with you joining our department. However, I do want to reiterate the importance of publishing. I'm sure you know this already – but I must state it again. You MUST publish or you WILL perish. So, publish, publish and publish more. OK? And within top tier journals with high impact factors.

STEFANIE: What I wanted to say was *"Wow! I had NO idea that was the expectation – thanks for telling me this important piece of news!"* But instead I said, "Yes. Thank you for sharing that. I am aware of the importance around publishing and I will try my best to live up to that expectation".

The concept of publishing is nothing new within academia and the constant pressure to produce manuscripts within peer-reviewed journals with high impact factors and Social Science Citation Index (SSCI) is the norm – following a capitalistic structure of productivity as "performance". However, there is, and has been, favouritism within our tourism studies discipline towards more positivist epistemological views that are inclusive of quantitative methodologies and frameworks. This paradigm is one of scientific production as a patriarchal one and that "academic leadership shows a clear hegemony of the affluent White man form a developed country" (Munar, 2016, p. 4). Early career researchers are still quite dependent on publishers, essentially for a symbolic reason: the award of "academic capital" and prestige while established researchers do the same to keep their research funding (Lariviere et al., 2015). As a critical scholar, Munar challenged what is considered as "right" or "valid" academic knowledge and posited that the research article is a genre with clear limitations and rules on "structure, size, submission, review, and publication processes and tools, copyright regulations, citation systems, and language" (2016, p. 14). More importantly, she argued for a critical discussion within tourism studies to rethink what it is to be an academic and how academic careers can be lived. However, as evidenced in the narrative above, this is challenging within departments or institutions that are not open to this critical discussion or alternative ways of producing in the academy.

The colonisation of tourism knowledge

Intertwined within the "publish or perish" paradigm of the academy is the production and colonisation of tourism knowledge (Munar, 2016). As scholars, we work to produce knowledge through collecting meaningful data and analysing said data using established theoretical lenses. In some cases, we produce new theories that expand the horizons of tourism studies. However, it is not often that we ask ourselves "who" the knowledge is being produced for and whether it is accessible to those outside the inner circle of the academy. In the extract below, we share a narrative with two industry professionals that highlights the disconnect between tourism as an academy and tourism as an industry. The scenario includes both of us in a conversation with two tourism professionals we are working with on a collaborative research project:

ALANA: We're really excited to work with you on this project to gain a better understanding of Black travel history! In fact, we've actually published a lot in this area already and shared the links to these articles with you last week.

EVELYN: This is great news! Honestly though, before meeting the two of you, I didn't even know tourism academics or research even existed. I've been wanting to do this type of research for so long, but with my other responsibilities I just didn't have the time, so I am so glad we connected.

UMAA: Yeah – I am also really happy we are connected. I actually saw the publications you mentioned on RESET's website, but I tried to access them and I can't get access without paying an exorbitant fee.

STEFANIE: I am sorry about that. It's academia. We do all the work, pay for the data collection, analysis and review papers and they still charge outsiders for access. Not to worry though, we will share the documents with you both directly using a shared folder.

About a month after this conversation occurred, Alana shared a recently published article on social media, attaching a full copy of the article using an online research platform Research Gate. In sharing, she received the following feedback from a fellow academic, Justin:

JUSTIN: Just a note worth considering – perhaps share the first page of the article only in Research Gate. That way you ensure the data in the journal's website is optimised (e.g. article views).

In response to this message, a conversation ensued between us about the ethical concerns surrounding this type of knowledge sharing. On the one hand, Justin is correct, in that directing people away from the journal website to read an article will alter the journal's website data on metrics. However, these journals are often not accessible without institutional access usually provided within the academy due to prohibitive access fees. Additionally, authors of journal publications are not financially compensated for their work beyond their institution's salary (e.g. most data collection and transcription software is paid for by the author/s), and they are also not compensated for time spent as a reviewer or journal editor. It certainly poses a moral and ethical dilemma when considering how tourism knowledge should be shared and who truly owns the intellectual property. As Munar eloquently posited:

> The evolution of knowledge and/or of science does not take place in closed communities separated from the rest of the world, but it is time and space contingent and deeply embedded in the larger processes of change experienced by contemporary societies.
>
> (2016, p. 5)

A tale of two pandemics

As we write this chapter, the world is in the middle of the COVID-19 global pandemic (World Health Organization, 2020). The far-reaching and deeply damaging impacts to the tourism industry have already become clear, exposing the crises and tensions steeped in inequity that have historically plagued our industry (Benjamin et al., 2020). However, as the pandemic continues, focusing only on the industry may cause us to gloss over the stark inequalities that are becoming increasingly more evident within academia. For instance, womxn's publishing rates have declined dramatically compared to men, and evidence shows that the pandemic is disproportionately affecting younger female researchers in particular (Viglione, 2020). These inequities are felt for female academics and researchers across disciplines, tourism being no exception. Womxn are also more likely to take responsibility for ailing relatives that leads to additional emotional labour (Viglione, 2020). Simultaneous to the health pandemic, there is also the pandemic of racial violence and injustice leading to social unrest in the USA (and elsewhere). At the University level, this has surfaced in the form of more committee work to address projects such as diversity, equity and inclusion strategic plans to include statements, goals and training across schools, departments and colleges. Additionally, the pivot to online spaces (i.e. Zoom) has forced academics into back-to-back meetings, since there is no requirement for travel and "assumed extra time" for more productivity. Some institutions also cancelled breaks (i.e. Fall or Spring Break) to adhere to a condensed, more "efficient" semester to limit students from traveling and potentially contracting COVID-19. We reflect on how these pandemics have impacted us in our professional lives:

ALANA: I don't know about you – but the pivot I have had to make since the pandemic began has really got me overwhelmed. Not only is it the extra time preparing classes in a virtual environment, but also – I have gotten so many more "asks" to serve on or lead committees, panels and workshops. It feels like it's all piling up at once.

STEFANIE: I am in the same boat! In addition to the already chaotic semester where we had to pivot online and re-design so much of our class design and pedagogy – I am on five committees/task forces around diversity, equity and inclusion, a 30% increase from last year. Being un-tenured as well – I know this is a heavy service load … but I feel it is necessary. And I also noticed, the majority of committee members across my University are womxn, Black and/or faculty of color!

ALANA: We have the same issue! I am actually leading the diversity planning committee for my department and every single member who volunteered is a womxn. Also, my University recently reported that across the board, the majority of committee members were womxn and many are untenured!

STEFANIE: In some ways, it is no surprise. It's long been known (and proven) that womxn and Black/faculty of color take on more service, especially around diversity and inclusion, than their counterparts. The pandemic has simply exposed this more clearly.

ALANA: You are right. And on top of all of that, we have both been tasked with so much more because of our work on marginalised populations in travel. In some ways it is nice to finally be recognised, but also overwhelming as I cannot afford to fall behind in publishing.

In addition to recent reports highlighting the inequities exposed by the COVID-19 pandemic, this narrative further humanises the impacts of the pandemic, and highlights it as a feminist issue specifically and an issue of invisible labour for Black and faculty of colour. This term "invisible labour" has been used to describe the unrecognised work that underrepresented faculty members are often called to do by virtue of their status (Flaherty, 2019). For example, this might include student mentoring and/or significant committee work, especially in the arena of diversity and inclusion. Most often, this work is taken on by those faculty more likely to identify as non-White, non-male or first-generation college students (Flaherty, 2019). However, reports continue to reveal that this type of "invisible labour" is not usually rewarded with tenure and promotion (Matthew, 2016). As such, it is imperative to create a system that fosters care and well-being for womxn and other traditionally marginalised faculty and values this type of service in evaluations for awarding tenure and promotion.

Discussion

Especially at this pivotal moment in history, the measure of our science as scholars of tourism needs to consider a "critical turn" in the direction of epistemologies and pedagogies that question the colonised and capitalist mindset of academia. This "critical turn" and its associated scholars "seek to stimulate their audience to transform society and thereby to liberate themselves and others" (Bramwell & Lane, 2014, p. 2). To begin the

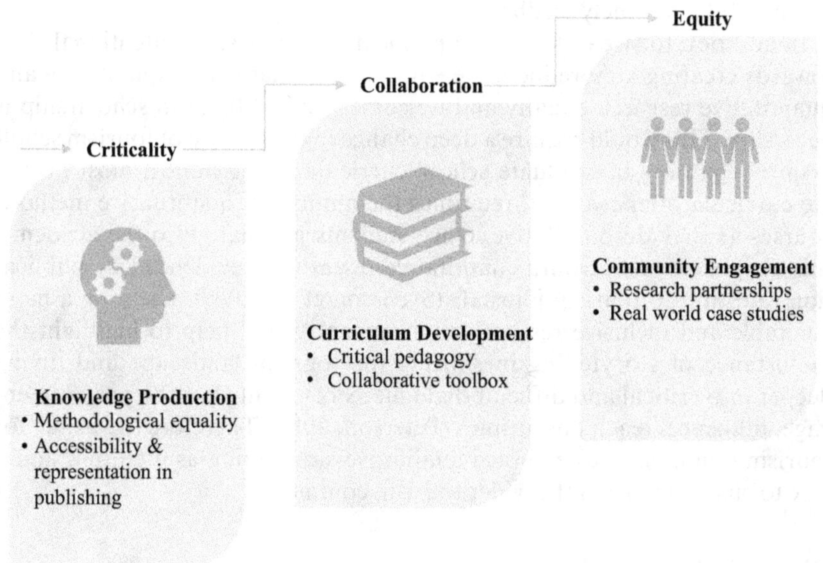

Figure 7.1 Blueprint to RESET tourism academia.

decolonisation process, we suggest a "blueprint" to address the systemic issues outlined in this chapter. Through the lenses of criticality, collaboration and equity, we suggest initiatives in the realms of knowledge production, curriculum development and community engagement (see Figure 7.1).

In the realm of knowledge production, we suggest increased accessibility and representation for both authors and readers. For authors, we recommend that academic journals provide a more streamlined process from submission to publication to include reasonable and transparent turnaround times. Additionally, journals must ensure that reviewers are qualified experts in both the topic area and methodology. For instance, encouraging or mandating implicit bias seminars can also help to reduce the potential discriminatory comments and even rejections from reviewers. This emerging body of cognitive and neural research helps to identify ways in which unconscious patterns people inevitably develop in their brains to organise information actually "affect individuals attitudes and actions, thus creating real-world implications, even though individuals may not even be aware that those biases exist within themselves" (Racial Equity Tools, 2020, paragraph 1). These embedded stereotypes and biases contribute to a system of inequity that perpetuates inequitable policies, behaviours and practices. We realise that this is a massive hurdle for many journals to standardise due to the voluntary nature of the review process. Therefore, we also suggest that journals introduce a modest compensation structure to assign value for reviewers' time and expertise. Streamlining the process in this way will trickle down to provide more clarity for authors wishing to publish and receive meaningful and timely feedback.

In addition to streamlining the publication process, an intentional shift towards creating a more inclusive environment that values qualitative and quantitative research equally and welcomes critical tourism scholarship is necessary. This would require a deep change in the culture of tourism scholarship beginning in graduate school curricula. As it stands, most graduate curricula offer two to three times the number of quantitative methods courses as they do qualitative. Once students graduate into the academy, this unspoken rule favours quantitative research as evidenced by publication rates in top tourism journals (Svensson et al., 2009). Creating a more equitable and inclusive research environment would help to highlight the importance of storytelling in shifting the tourism landscape and diving deeper into critical and difficult dialogue. Storytelling has the power to engage, influence, teach and inspire (Peterson, 2017). Therefore, we argue for tourism scholarship to embrace qualitative storytelling as a culture and a way to enact change in the wider tourism community.

RESET-ing how knowledge is circulated

Through Tourism RESET, we are striving to reach goals which bridge academia and community engagement – creating a holistic paradigm and approach of knowledge production. For instance, we created pedagogical and research toolboxes, free and open to the public, to assist with disseminating information around critical tourism scholarship and pedagogy. Our hope is that amplifying counter-narratives and voices of marginalised populations will help to contribute to an anti-racist and more holistic comprehension of tourism studies. Furthermore, we are posting social media content and blog posts that are easy to digest and share that illuminate relevant topics and concerns for various audiences and populations. Although these forms of *publications* may not be as *valuable* to our prospective institutions or assist with tenure or promotion, we feel it is important and essential to have these dialogues outside the siloed academic landscape. Especially with the impact of COVID-19, many academics are reimagining new ways to connect with learners and the broader public. As Connolly argued, “scholarship requires expertise, confidence, trust, and love of a better world” (2020, paragraph 15). As such, we propose a paradigm shift which advocates for a more flexible approach to defining what counts in the production and circulation of knowledge.

Engaging our communities beyond institutionalised landscapes is also part of RESET’s vision. We are working on several collaborations with organisations such as the Black Travel Alliance and NOMADNESS Travel Tribe, and recommend this type of collaboration to the wider network of tourism scholars. These partnerships not only assist with helping our industry move towards an equitable landscape, but also illuminate the important work of tourism scholarship, bringing public attention to its significance and impact. By creating relationships outside of academia, we can work towards dismantling the traditional hegemonic perception of the academy. In order to RESET, we must rethink what "counts" as productivity and work towards redesigning how academia chooses to reward the production and circulation of knowledge. By breaking down these barriers of access and academic jargon, it is our hope that we can instil a respect for the academy and help to advocate for policy reform, equitable landscapes and provide critical educational material for academic institutions to be accessible to all. Hopefully this dismantlement of the stereotypical ivory tower will contribute towards the socialising of what productive and valuable tourism research *can* look like – creating a revaluation of tourism academia.

Notes

1 The term womxn is an inclusive term that is used, especially in intersectional feminism, as an alternative spelling to "woman" to avoid the suggestion of sexism and to be inclusive of trans and non-binary women.

2 We would like to note that we are aware of the paradoxical use of scholars like Tribe and Bramwell and Lane to justify the diversification of voices. It is frequently these White male scholars of influence who do "state of the art"/ "meta-analysis", building on the work of those who do detailed qualitative work, which become highly referenced and thus contributing to the very problem they are trying to resolve.

3 Research I university is a category that the Carnegie Classification of Institutions of Higher Education uses to indicate universities in the US that engage in the highest levels of research activity.

References

Arday, J. (2018). Dismantling power and privilege through reflexivity: Negotiating normative Whiteness, the Eurocentric curriculum and racial micro-aggressions within the Academy. *Whiteness and Education, 3*(2), 141–61.

Ateljevic, I., Pritchard, A., & Morgan, M. (2007). *The critical turn in tourism studies: Innovative research methodologies*. Amsterdam: Elsevier.

Benjamin, S., Dillette, A. & Alderman, D.H. (2020). "We can't return to normal": Committing to tourism equity in the post-pandemic age. *Tourism Geographies, 22*(3), 476–483. DOI: 10.1080/14616688.2020.1759130

Bianchi, R. V. (2009). The "critical turn" in tourism studies: A radical critique. *Tourism Geographies, 11*(4), 484–504.

Bramwell, B., & Lane, B. (2014). The "critical turn" and its implications for sustainable tourism research. *Journal of Sustainable Tourism, 22*(1), 1–8.

Cannella, G. S., & Lincoln, Y. S. (2017). Ethics, research regulations, and critical social science. In N.K. Denzin & Y.S. Lincoln (Eds.), *The Sage handbook of qualitative research* (pp. 75–90). Los Angeles, CA: Sage Publications

Caton, K. (2015). Studying tourism: Where's the humanity in it? In P. J. Sheldon & C. H. C. Hsu (Eds.), *Tourism education: Global issues and trends* (pp. 15–39). Bingley: Emerald Group Publishing.

Connolly, J. (2020, April 9). Inside higher ed. Retrieved 7 December 2020, from https://www.insidehighered.com/advice/2020/04/09/covid-19-demands-reconsideration-tenure-requirements-going-forward-opinion

Dillette, A. K., Benjamin, S., & Carpenter, C. (2019). Tweeting the black travel experience: Social media counternarrative stories as innovative insight on #TravelingWhileBlack. *Journal of Travel Research, 58*(8), 1357–72.

Ellis, C. (2004). *The ethnographic I: A methodological novel about autoethnography.* Walnut Creek, CA: AltaMira Press.

Ellis, C., & Bochner, A. P. (Eds.). (1996). *Composing ethnography: Alternative forms of qualitative writing.* Walnut Creek, CA: AltaMira Press.

Ellis, C., & Flaherty, M. (Eds.). (1992). *Investigating subjectivity.* Newbury Park, CA: Sage.

Ferguson, E. (2020, June 16). Guest column: A travel industry discussion of race and racism. Retrieved 7 December 2020, from https://www.ustravel.org/news/guest-column-travel-industry-discussion-race-and-racism

Flaherty, C. (2019, June 4). Who's doing the heavy lifting in terms of diversity and inclusion work? Retrieved 7 December 2020, from https://www.insidehighered.com/news/2019/06/04/whos-doing-heavy-lifting-terms-diversity-and-inclusion-work

Goodall, H. L., Jr. (2000). *Writing the new ethnography.* Lanham, MD: AltaMira.

Gubrium, J., & Holstein, J. (2008). Narrative ethnography. In S.N. Hesse-Biber & P. Leavy (Eds.), *Handbook of emergent methods* (pp. 241–64). New York: Guilford Publications.

Habermas, J. (1978). *Knowledge and human interests.* London: Heinemann.

Hertz, R. (1997). *Reflectivity and voice.* Thousand Oaks, CA: Sage.

Higgins-Desbiolles, F. (2006). More Than an "Industry": The forgotten power of tourism as a social force. *Tourism Management, 27*, 1192–208.

Kwan, M-P & Schwanen, T. (2009). Quantitative revolution 2: The critical (re)turn. *The Professional Geographer, 61*(3), 283–91.

Lariviere, V., Haustein, S., & Mongeon, P. (2015). Big publishers, bigger profits: How the scholarly community lost the control of its journals. *Media Tropes, V*(2), 102–10.

Lyotard, J. F. (1984). *The postmodern condition: A report on knowledge* (Vol. 10). Minneapolis: University of Minnesota Press.

Matthew, P. (2016, November 23). What is faculty diversity worth to a university? Retrieved 7 December 2020, from https://www.theatlantic.com/education/archive/2016/11/what-is-faculty-diversity-worth-to-a-university/508334/

Munar, A. M. (2016). The house of tourism studies and the systemic paradigm. In A.M. Munar & T. Jamal (Eds.), *Tourism research paradigms: Critical and emergent knowledges* (Vol. 22, pp. 131–53). Bingley: Emerald Group Publishing Limited.

Peterson, L. (2017, October 17). The science behind the art of storytelling. Retrieved 7 December 2020, from https://www.harvardbusiness.org/the-science-behind-the-art-of-storytelling/

Pritchard, A., & Morgan, N. (2007). De-centering tourism's intellectual universe, or traversing the dialogue between change and tradition. In I. Ateljevic, A. Pritchard, & N. Morgan (Eds.), *The critical turn in tourism studies: Innovative research methods* (pp. 11–28). Amsterdam: Elsevier.

Racial Equity Tools (2020). Retrieved 7 December 2020, from https://www.racialequitytools.org/act/communicating/implicit-bias

Riley, R. W., & Love, L. (2000). The state of qualitative tourism research. *Annals of Tourism Research, 27*(1), 164–87.

Spinks, R. (2020, June 26). Tourism marketers must now go beyond optics to reach black travelers. Retrieved 7 December 2020, from https://skift.com/2020/06/26/tourism-marketers-must-now-go-beyond-optics-to-reach-black-travelers/

Svensson, G., Svaeri, S., & Einarsen, K. (2009). "Empirical characteristics" of scholarly journals in hospitality and tourism research: An assessment. *International Journal of Hospitality Management, 28*(3), 479–83.

Tribe, J. (1997). The indiscipline of tourism. *Annals of Tourism Research, 24*(3), 638–57.

Tribe, J. (2008). Tourism: A critical business. *Journal of Travel Research, 46*, 245–55.

Viglione, G. (2020, May 20). Are women publishing less during the pandemic? Here's what the data say. Retrieved 7 December 2020, from https://www.nature.com/articles/d41586-020-01294-9

Walle, A. H. (1997). Quantitative versus qualitative tourism research. *Annals of Tourism Research, 24*(3), 524–36.

Worland, J. (2020, June 11). America's long overdue awakening on systemic racism. Retrieved 7 December 2020, from https://time.com/5851855/systemic-racism-america/

World Health Organization (2020). Retrieved 7 December 2020, from https://www.who.int/

Xiao, H., & Smith, S. (2006). The making of tourism research: Insights from a social sciences journal. *Annals of Tourism Research, 33*(2), 490–507.

8 DeTouring the empire

Unsettling sites and sights of US militarism and settler colonialism in Hawai‘i

Kyle Kajihiro

Introduction

On a bright December morning in 2011, shortly before the 70th anniversary of the bombing of Pearl Harbor, I helped to lead a field trip for 57 Honolulu high school students to the Pearl Harbor National Memorial. I wanted to contextualise their study of the Second World War by conveying a sense of the rich and complicated history of Ke Awalau o Pu‘uloa – the Hawaiian name for the place where Joint Base Pearl Harbor Hickam is located – once *‘āina momona* (abundant and life-giving place) for *Kānaka ‘Ōiwi* (Native Hawaiians) and still an object of US imperial desire. "Āina", which means "that which feeds", refers to land, but also invokes a genealogical relationship between humans and the earth.

Our students were from Kalihi, a largely immigrant working-class neighbourhood in Honolulu. Their ethnicities – Filipino/a, Sāmoan, Chinese, Vietnamese, Chuukese, Marshallese, Japanese and Native Hawaiian – sketched a map of war and imperialism in the Pacific. After sharing Kanaka ‘Ōiwi place names and stories, and discussing how the attack on Pearl Harbor can be considered an outcome of the competing imperialisms of Japan and the United States, we asked them to consider why the line "The Kingdom of Hawai‘i was overthrown in 1893" would be considered the most controversial text in the entire exhibit and how this memorial might look different if it was dedicated to peace rather than war. We examined how martial law in Hawai‘i and the internment of persons of Japanese ancestry related to anti-Muslim policies instituted after Al-Qaeda's attacks on the United States on 11 September 2001.[1]

On the lawn, large tents and chairs were set up for the upcoming Pearl Harbor commemoration ceremonies. We joined a group of visitors in the shade. But as we sat, a sailor in blue camouflage told us that we were not allowed to sit on the chairs. Noticing that the other visitors were allowed to remain seated, I asked the sailor if he was going to enforce the "no-sitting" rule with them as well.

His face grew red. He leaned forward and barked, "What's your name?! Who are you with?! Are you telling us how to do our job?"! These verbal

DOI: 10.4324/9781003164616-8

warning shots were meant to intimidate and put me in my place. I pointed out that his actions were discriminatory. He grunted that he would talk to the other visitors, "when I get around to it". As I walked away, he griped to his partner, "Fucking bitch"!

The youth, who had witnessed the whole exchange, were abuzz. I told them to pay attention to how we were treated, to who was allowed to sit and who was not. I asked them to reflect on why we were treated this way. Several blurted out, "It's racism, mister"! "They only care about tourists"! This small confrontation with military power became an excellent teachable moment.

After returning to campus, the students developed short skits about what they had learned. Several groups did parody re-enactments of our encounter with the sailors at Pearl Harbor. One student-actor, summoning the full authoritarian absurdity of the scene, bellowed, "Can't you see how straight these chairs are?"! which sparked peals of laughter. While there is more here to unpack, I offer this anecdote to illustrate the kinds of affectively charged and pedagogically rich encounters that sometimes occur on the educational programme we call Hawai'i DeTours (Kajihiro, 2011).

This chapter examines the promise and the limitations of the Hawai'i DeTours Project, a critical educational programme which promotes a decolonial history and geography of Hawai'i. If the military in Hawai'i is everywhere and yet "hidden in plain sight" (Ferguson & Turnbull, 1999, p. xiii), then I ask whether critical educational tours of this "lost geography" (Smith, 2004) can effectively unsettle and challenge militarism in the islands. However, as the Hawai'i DeTours Project grows in popularity, interest from visitors who want a more authentic or guilt-free experience has grown. If the Hawai'i DeTours Project were to expand its offerings to fee-paying customers, would it risk becoming just another "socially responsible" commodity? Or if we refuse to meet rising demand, do we fade into obscurity? Can meaningful transnational solidarity be fostered through these kinds of practices? This chapter gives an overview of the history and praxis of the Hawai'i DeTours Project and offers some tentative reflections on these questions, including its successes and challenges.

I argue that decolonial educational tours can offer an effective method of political education and organising support for local struggles. However, as Haunani-Kay Trask writes, "To most Americans, ... Hawai'i is theirs: to use, to take, and, above all, to fantasize about long after the experience" (1993, p. 180). However, given the degree to which "Hawaii" is overdetermined as a militouristic commodity, it is difficult under present conditions for Hawai'i DeTours to scale up its operations significantly without becoming another novelty for touristic consumption. If tourism and militarism within capitalist and colonial relations are inherently extractive and violent, then a more "woke" (hip and socially aware) tourism can never be a real alternative. The challenge, then, is to develop a different ethical relation to travel and mobility which is grounded in decolonial politics of solidarity and *kuleana* (privilege and responsibility) to peoples and places. The Kanaka ʻŌiwi concept of *huakaʻi* (a journey with purpose, entailing relationships and mutual responsibilities between traveller and resident) suggests ways to remake our orientation to travel based on social and ecological values of reciprocity, care and social justice.

Unsettling the militourism complex

The late Banaban and African American scholar Teresia Teaiwa defines "militourism" as "a phenomenon by which military or paramilitary force ensures the smooth running of a tourist industry, and that same tourist industry masks the military force behind it" (Teaiwa, 1999, p. 251). Vernadette Gonzalez extends the concept to interrogate the articulation of militarism and tourism to reproduce settler colonial and neocolonial formations in Hawaiʻi and the Philippines (Gonzalez, 2013).

Military bases have long provided the force by which to extend the United States' imperial reach across the Pacific to maintain geopolitical hegemony and secure conditions for unfettered capitalist accumulation (Davis, 2015, 2017; Smith, 2004; Vine, 2015). From the mid-1800s, Hawaiʻi was a fulcrum in the process of American imperial formation during its crucial pivot from a continental to maritime empire. Despite Hawaiʻi's importance to the US imperial project, it remained a "lost geography" (Smith, 2004) of American empire, largely overlooked as a problem deserving more critical attention because of the thorough naturalisation of US settler colonialism on the islands. This "closing of the frontier" in Hawaiʻi produces a blind spot on the map which conceals the United States' unstable and contested claims in Hawaiʻi. From this blind spot, the United States is able to coordinate military activities and bases throughout the vast Indo-Pacific Command area of responsibility, which extends from the west coast of North America to the central Indian Ocean and from Alaska to the entire continent of Antarctica (US Indo-Pacific Command, 2020).

Tourism in Hawaiʻi has always constituted a crucial component of this US imperial project (Feeney, 2009; Imada, 2004, 2012; Skwiot, 2010; Thompson, 2010). Creating the "Paradise of the Pacific" was as much an ideological and geopolitical undertaking as a commercial venture to develop "one of the most successful tourist destinations in the world" (Mak, 2015, p. 1). Touristic discourses about Hawaiʻi serve to make the islands more attractive and assimilable by projecting idyllic images which conceal and naturalise the structural violence of settler colonialism and militarism, smoothing over contradictions and potentially disruptive differences. Many relate to Hawaiʻi through fantasy conceptions of tropical paradise or multicultural melting pot – somewhere distant and exotic enough to excite consumer desires, but familiar enough so as to not discomfort or unsettle visitors. Because of the ubiquity of touristic discourses about Hawai'i, everyone is always already placed in some kind of relationship to Hawai'i mediated by tourism and militarism. The militourism complex has effectively produced a "Hawaii" *for* America. Projects that engage in the unlearning of these naturalised (mis)understandings about Hawai'i provide a ripe opportunity for imagining and creating alternative futures.

History of the Hawai'i DeTours Project

From 1996 to 2011, I served on the programme staff of the American Friends Service Committee (AFSC) Hawai'i Area Programme, a peace and social justice organisation. As a fourth-generation settler of Japanese ancestry in Hawai'i whose family members have traversed the imperial terrain of the Pacific, my programme work focused on countering the destructive impacts of militarisation in Hawai'i. Our organisation often hosted social justice activists and scholars from around the world. We were perplexed and frustrated that so many smart and politically conscientious people switched off something in their brains when they heard "Hawaii" and transformed into giddy teenagers in a tropical playground (Neason, 2020). Life-long activists who protested against every US war from Vietnam to Afghanistan and Iraq suddenly forgot that such wars were made possible by the United States' vast "carceral archipelago" (Foucault, 1979) of militarised sites, logistical infrastructure and colonised places. In order to highlight the connections between militarisation in Hawai'i and the wars rehearsed on and deployed from Hawaiian lands, our programme organised a series of geopolitical reality tours of military-occupied and contested sites on O'ahu for our members and visiting groups. I also began to lead informal tours with visiting activists and scholars.

In the late-1990s, I began working with Terri Keko'olani, a veteran Kanaka 'Ōiwi activist who participated in many struggles, including the successful movement to stop the Navy bombing of Kaho'olawe island. In 1999, we travelled to Vieques, Puerto Rico, on a solidarity delegation sponsored by the Fellowship of Reconciliation and Proyecto Caribeño para Justicia y Paz to support the growing movement to oust the US Navy. These activist exchanges brought people together in solidarity at sites of struggle. By making the costs and consequences of militarisation in Hawai'i more visible and understandable, we aimed to explore alternative futures and build wider solidarity networks. Between 2000 and 2011, our programme engaged in a number of campaigns to oppose military expansion and to recover Hawaiian lands from the US military. The DeTours became a valuable vehicle for political education about these issues.

Terri and I co-founded the Hawai'i DeTours Project to unsettle participants' preconceptions of Hawai'i with critical information about Hawaiian history, (re)kindle their connections to places through stories of the land, connect them with *kia'i* (protectors of the land) who work for creative alternatives and foster stronger political commitments and relationships in support of ongoing campaigns. We challenge DeTour participants to wrestle with their own *kuleana*. And these kuleana would vary depending on whether they were Kānaka ʻŌiwi, local residents or various differently situated visitors. What responsibilities do they have to the peoples and places they encountered, to the new knowledge, experiences and relationships they have gained?

In 2011, the Hawai'i DeTours Project became a programme of Hawai'i Peace and Justice, the successor to the AFSC Hawai'i Area Program. Growing mainly by word of mouth, to date we have hosted approximately 1,500 persons, of whom the majority are Hawai'i residents. The tours have ranged from a single passenger in my car to several hundred persons on multiple buses.

We originally called the tours "demilitarisation tours" or "decolonial tours", but shortened it to "detours", a journey that swerves off the beaten path. The name is also a play on the Situationists' concept of *détournement*, an overturning or hijacking of a work of art to subvert its meaning or to make a political statement (Debord & Wolman, 1956). Our DeTours can be thought of as an appropriation and negation of the commercial group tour in order to advance decolonial aims.

The itinerary

In the beginning, our itinerary took an entire day and typically included visits to rural sites where communities were fighting to reclaim land from the military. As we got more requests to lead DeTours for university groups and conferences, we narrowed our focus to a few key sites related to the history of the US in Hawai'i. Our first stop on the DeTour is the 'Iolani Palace, a living symbol of Hawaiian sovereignty and the "scene of the crime", where in 1893, the US military backed a *haole* (white settler) *coup d'état*. It was America's first regime change of a sovereign state (Kinzer, 2006). Here, our discussion focuses on the history of the Hawaiian Kingdom and the complex political-economic and cultural changes underway in 19th century Hawai'i. Haole sugar planters pushed King Kalākaua to negotiate a treaty of reciprocity with the United States which would remove tariffs on Hawai'i-grown sugar. Meanwhile, US military planners sought a naval coaling station at Ke Awalau o Pu'uloa. These two interests propelled a chain of events leading to the settler plot to overthrow Queen Lili'uokalani in pursuit of Hawai'i's annexation to the United States.

At the front gate of the Palace, near the spot where the US Marines posted their Gatling guns during the overthrow, Terri holds up a photo of her *kupuna* (elder or ancestor) Solomon Peleioholani, a historian in the Hawaiian Kingdom. Between 1893 and 1898, Hawaiian nationals organised a mass movement to resist annexation. In 1897, more than 90% of Hawaiian nationals signed two anti-annexation petitions which helped to defeat a treaty of annexation in the US Senate. Without such a treaty, the legality of the United States' annexation of Hawai'i remains clouded ("Kue: The Hui Aloha Aina Anti-Annexation Petitions, 1897–1898", 1897). Today, the Hawaiian Kingdom flag flies alone above the 'Iolani Palace, symbolizing the living sovereignty of the land; it is one of two public buildings where no American flag flies over the Hawaiian flag.

Terri asks a tour participant to read aloud the text of the Kū'ē (resistance) Petition, which protests annexation to the United States "in any form or shape" ("Kue: The Hui Aloha Aina Anti-Annexation Petitions, 1897–1898", 1897). She points out her kupuna's signature on the page and shares how she found his name in the petition in 1998 when the documents were publicly displayed in Hawai'i for the first time in 100 years.

On the palace grounds, we describe key developments in the modern Hawaiian sovereignty movement, including the 1993 'Onipa'a march to 'Iolani Palace commemorating the 100th anniversary of the overthrow. At the *ahu* (stone altar) in the corner of the palace grounds, we recount how Kānaka 'Ōiwi brought stones from their homelands across the archipelago to build that structure, symbolising the *lāhui* (nation) united. While it is a modern structure, this ahu has become a sacred site through the ceremonies and events investing it with *mana* (power). Some share their own family stories about the overthrow and resistance to annexation. Others reflect on why they had never learned this history in school.

On Hālawa Heights, overlooking Ke Awalau o Pu'uloa, we stop outside Camp Smith, headquarters of the United States Indo-Pacific Command (USINDOPACOM). In 1873, General John Schofield led a secret scouting mission to the Hawaiian Islands posing as tourists. I ask the group to imagine Schofield coming upon a view like ours and concluding that this was "the key to the Central Pacific Ocean, the gem of these islands" (Schofield, 1873).

The US Indo-Pacific Command was established in Hawai'i in 1947. It is the oldest and largest of the unified military commands. Borrowing a metaphor from Kanaka 'Ōiwi scholar-activist Kaleikoa Ka'eo, I explain how the US military in Hawai'i is like the head of a monstrous *he'e* (octopus), with a "brain" and "nervous system" (command centres, supercomputers and communication networks), "eyes" (radar facilities and optical tracking stations) on sacred mountain tops, "ears" (signals intelligence and radio communications facilities), "excrement" (toxic contamination and other negative externalities) and "tentacles" (the network of bases) extending across the Pacific gripping other places like Kwajalein, Okinawa, Guåhan (Guam), Korea, Japan and the Philippines.

The DeTour continues on through the military landscape: neatly manicured housing areas, sprawling golf courses and shopping complexes. I point out contrasts between civilian and military areas and explain the inequalities produced by the military political economy, such as housing costs inflated by housing subsidies for military personnel which financially squeeze non-military residents.

Our next stop is the Pearl Harbor National Memorial, Hawai'i's most visited tourist attraction, where we tell the story of how Ke Awalau o Pu'uloa became "Pearl Harbor", a site of US military power and national memory. We remind the group that many consider Ke Awalau o Pu'uloa/Pearl Harbor to be a sacred place, a site of Native Hawaiian religious significance as well as an American war memorial. We ask them to be critical, yet respectful. By stretching the temporal scope and the boundaries of sacredness, we deprivilege the US narrative of Pearl Harbor and situate it within a longer *mo'okū'auhau* (genealogy) of Hawai'i (Mei-Singh & Gonzalez, 2017).

At the large floor map of the Pacific region we consider the implications of different geographical imaginaries, whether the Pacific region is a sparsely populated "American Lake" or a magnificent "sea of islands" full of complex genealogical, cultural and ecological relationships (Hau'ofa, 1994). With Ke Awalau o Pu'uloa as a backdrop, we share *wahi pana* (storied places) and *mo'olelo* (historical accounts). In one mo'olelo, Ka'ahupāhau, the great shark goddess of 'Ewa, banned the killing of humans and ordered local sharks to protect the people from invaders. Another mo'olelo explains how Kānekua'ana, a *mo'o wahine* (a female reptilian guardian deity), provided and protected the abundant marine resources of Ke Awalau o Pu'uloa until an abusive and irresponsible chief caused her to revoke her gift. The construction of Pearl Harbor destroyed a complex eco-social system based on the abundant fresh water of the district: *lo'i kalo* (taro pondfields) and *loko i'a* (walled fishponds) were filled in and paved, reefs teeming with fish and shellfish dredged and Indigenous tenants displaced.

In the museum exhibition, we critically examine the historical narrative: the multiculturalisation of the Second World War memory; how the US and Japanese versions of events are presented symmetrically, reflecting the postwar rehabilitation of Japan from enemy to anti-communist ally in East Asia; how the Japanese American experience of the war is framed by the racism of internment camps and redemption through war heroism. We ask the participants to notice how the display depicts casualties of the attack on Pearl Harbor in graphic detail while representing the atomic bombing of Hiroshima with a single photo of the devastated city devoid of human casualties, and to think about what this says about whose lives are considered grievable. We ask them to consider how the placement of the "Native Hawaiian" story on a series of placards in the breezeway outside of the main exhibition, where people typically sit with their backs to the displays, exemplifies the settler colonial erasure and simultaneous enlistment of Indigenous presence.

Our last stop is Hanakehau Farm, near the shores of Ke Awalau o Pu'uloa in Waiawa, where Kanaka 'Ōiwi activists are restoring spring-fed lo'i kalo and conducting educational programmes in Kanaka 'Ōiwi cultural practices and trainings in non-violent, direct action methods. Andre Perez and Camille Kalama share the history of the 'āina and their work and discuss their mission to grow Hawaiian consciousness. As we eat our lunches an arm's throw away from the Navy's toxic dump site, we debrief the day's experiences and discuss possible decolonial futures for Ke Awalau o Puʻuloa, for Hawaiʻi and for the Pacific (Kajihiro & Kekoʻolani-Raymond, 2019).

DeTours as decolonial place-making

The Hawaiʻi DeTours Project has been an effective political educational practice for movement building by incorporating critical historical geography, counter-place making practices, building relationships with local activists and a vision of alternative futures. We believe that the DeTours have helped to normalise critiques of militarisation in the political discourse of social justice movements in Hawaiʻi and beyond.

Counter-place making in Hawaiʻi DeTours consists of several elements. First, we centre Indigenous geographies and try to animate the moʻolelo ʻāina (history of the land). By highlighting Kanaka ʻŌiwi place names, stories and historical events, we aim to change the way participants relate to places. By telling and retelling moʻolelo *in place*, we engage in a kind of *performance mapping* (Louis, 2017; Oliveira, 2014), where the recitation of stories, songs, dance and theatre summon the relationship with place. The land itself becomes a kind of archive or text to hold the stories.

Second, by critically interrogating familiar places, we seek to unsettle the military settler landscape. In other words, by making the familiar strange we aim to decentre naturalised understandings about places. Place is a kind of palimpsest, where layers of historical changes are overwritten, but never completely erased. Thus, the past can haunt and erupt into the present, forcing reckonings with difficult histories. On our DeTours, we make openings for these difficult pasts to unsettle the present.

Laurel Mei-Singh and Vernadette Gonzalez describe DeTours as a genealogical methodology, both in the Foucauldian sense of unearthing and interrogating accepted notions of truth and in the Kanaka ʻŌiwi sense of moʻokūʻauhau, a mapping of a web of relationships to peoples, places and events through time and across space (2017, p. 175). They argue that "DeTours interrupt colonial projects to convey the overlapping processes that shape Hawaiʻi's landscape while highlighting Kanaka relationships to *ʻaina* [sic]" (Mei-Singh & Gonzalez, 2017, p. 176).

Another aspect of unsettling the settler landscape is to expose the violence of occupation that may be hidden in plain sight. We once took a group of visiting South Asian scholars on the DeTour. After a brief stop at Camp Smith, we were detained by military police who demanded to view and possibly censor our photographs and video, despite the fact that we were on a public street and had not broken any laws. After a few tense moments of politely refusing to comply, we were allowed to leave. I joked with our shaken guests that they were lucky to have gotten the "VIP treatment". Such affectively charged moments can reveal the degree to which militarisation is inherently insecure and unstable and must continually be reproduced through performative acts of force.

Third, we seek to create new relationships between participants and the peoples and places we visit. As much as possible we try to meet with the local *kia'i* or protectors. It is critical that we share stories of resistance, (re) emergence and hope so that participants do not leave with a sense of political paralysis or cynicism.

Sometimes we engage in small actions to hold space within settler places, even if only momentarily, by invoking Kanaka 'Ōiwi ceremonies or other creative acts to disrupt normal activities. Once we took a group of Kanaka 'Ōiwi high school students to the Pearl Harbor memorial, where we asked them to analyse the museum's historical narratives. One boy sat off to the side looking dejected. I asked him what was wrong. He said, "I feel junk. Like I don't belong here. This doesn't feel like our place". Despite the rich 'Ōiwi history of Ke Awalau o Pu'uloa, these youth felt utterly out of place, alienated by the militouristic erasure of the Indigenous geography.

I asked what we could do to change the feeling of the space. He suggested doing an *oli* (chant). Forming a circle on the lawn, surrounded by pale tourists, the youth chanted a powerful oli to the ancestors and deities of the land. In the middle of this venerated US war memorial, a space opened up around us, a temporal and spatial *kīpuka* – an opening or oasis of Kanaka 'Ōiwi political and cultural difference. Visitors stared, but kept their distance. Park rangers seemed perplexed. This tiny act of creative defiance showed the students that their words and actions have power to transform space.

Geographers understand place to be produced relationally. Place is a process, a constant becoming. And each DeTour is a process of forming new relationships between people and the environment. In this way, place is produced through the stories we tell, the experiences we share, the memories we carry forward and the shifts in consciousness and political commitment we are able to inspire. In settler colonial situations, producing a new sense of place is a necessary step towards the recovery and transformation of these places.

Weaving 'upena of solidarity

Place is also produced through "a particular constellation of social relations, meeting and weaving together at a particular locus" (Massey, 1991, p. 28). One of our main objectives is to bring together people from different parts of the world to build support for struggles in Hawai'i and to broaden the political and cultural networks of local groups. We have hosted numerous activists and delegations from other countries engaged in similar struggles, including Guåhan (Guam), Puerto Rico, Japan, South Korea, the Philippines, Okinawa, Australia, Aotearoa New Zealand, the Marshall Islands, Germany, Canada, numerous Indigenous Nations in North America and various parts of the United States. These richly generative encounters have sometimes led to ongoing collaborations. One of our most successful relationships is with the Okinawan peace movement. Over the years, we have hosted at least four activist delegations from Okinawa. The DeTours have influenced the political rhetoric of the peace movement organisations in Okinawa. At one time in the 1990s, Okinawan anti-base leaders called for US military bases to be moved from Okinawa to "America", by which they meant either Guam or Hawai'i. However, after participating in our DeTours, these activists changed their rhetoric. Some even struggled with their own groups to stop demanding that the bases go to Hawai'i. Activists in Hawai'i also visited Okinawa and organised Okinawa solidarity actions in Hawai'i. We were able to cultivate a "politics of affinity" (Davis, 2015) by weaving an *'upena* (net) of *pilina* (relationships or ties) (Osorio, 2018).

Challenges and limitations

While our DeTours have raised critical awareness about the US military impacts in Hawai'i for a significant number of participants, its growing popularity has revealed contradictions with the project. One of the main challenges facing Hawai'i DeTours is problems of scale and capacity (Mzezewa, 2020). Despite its impacts on activism, the programme may reach a tiny fraction of the millions of annual visitors to Hawai'i and Pearl Harbor. In order to make a bigger impact, the programme would need to expand its operations considerably. Some have suggested offering DeTours to paying customers as a political education programme and revenue-generating activity, but we have been reluctant to do so, for reasons I discuss below.

Most of our group DeTours are planned collaboratively and intentionally with a partner. Were we to shift to a more systematised and business-like operation, we would risk losing the collegiality, immediacy and creativity of activist collaborations. The DeTour is a journey together, each one unique and drawing upon the knowledge and experience of the participants.

A small team helps to lead the DeTours. We have begun training other activists who are interested in becoming guides. Others have asked to learn our content hoping to commercialise it themselves. We have declined those requests.

The coronavirus pandemic's crippling of Hawai'i's tourism industry has stimulated critical reflection on the social and environmental impacts of tourism and sparked important discussions about stronger controls and limits on tourism in line with the principles of social and environmental justice ('Āina Aloha Economic Futures, 2020; Hawai'i State Commission on the Status of Women, 2020). We have also reflected on the ethics of offering DeTours to would-be visitors if it might be seen as giving permission or invitation to travel to Hawai'i by certifying their visit as more socially responsible (Neason, 2020). This sparked thinking about creating online multimedia content so that people could virtually participate without having to travel to Hawai'i.

The financial sustainability of the project is another limitation. For many years we led the tours as volunteers, free of charge. But as the tours grew in size and quantity, we started to request honoraria from institutions to cover the costs for ourselves and our community partners. The increase in inquiries made us consider offering regularly scheduled paid DeTours, both as an educational programme and a revenue generator for Hawai'i Peace and Justice. But would a fantasy-spoiling tour like ours ever be feasible? Indeed, a group of college business students prepared a business plan for DeTours as a class project and concluded that it would not be a viable business model.

However, our intention was never to be a business. Commercialising the DeTours raises more profound political and ethical questions. How does charging a fee alter the nature of the tour? While marketing our DeTours may enable us to reach larger audiences, the DeTour might become something entirely different if paid visitors participated, not to stand in solidarity, but to have a novel experience or absolution for their indulgence. The overdetermination of "Hawaii" as a tourist imaginary helps to explain why a social justice warrior might suddenly don a "tacky Hawaiian shirt state of the mind" at the thought of a visit.

The value of the Hawai'i DeTours derives from the relationships and mutual learning produced by the collective labour of tour leaders, partner organisations, community hosts and participants alike in relation to the 'āina itself. These relations would change profoundly if DeTours were to take a "leap of faith" across the irreducible gap from use value to exchange value in becoming a commodity (Žižek, 2006).

To illustrate the contradictions of solidarity tourism, I turn to our one ill-fated attempt to offer the DeTour as a tour package. This incident highlights why even activist-oriented tourism cannot advance meaningful social change in Hawai'i at this time. A few years ago, Global Exchange, an organisation that sponsors social justice-oriented Reality Tours to Cuba and other destinations, asked if we would help to organise a Hawai'i Reality Tour based on our DeTours. Modelled on solidarity delegations, our itinerary aimed to provide immersive and transformative experiences and deep, productive engagement with activists and community members.

Jeanne Cooper, a travel columnist for the *San Francisco Chronicle*, whose "Hawaii Insider" column trades on her supposed insider status as a part-time Hawai'i resident, wrote a scathing review of the Hawai'i Reality Tour itinerary:

> Loss of indigenous sovereignty? Check. Militarization? Check. Cultural and environmental degradation? Well, you get the picture. But do you really need to join a tour group to discover those pressing concerns . . . or to have meaningful encounters with island residents working to reverse or mitigate them?
>
> (Cooper, 2011)

Cooper ripped apart the itinerary point-by-point and snarkily offered her own "suggestions for do-it-yourselfers (and think-for-yourselfers)" (Cooper, 2011). For Cooper, touring social and environmental problems and having "meaningful encounters with island residents" are just items on her must-do list for "think-for-yourselfers". This approach perfectly illustrates the individualistic, consumer ideology of contemporary *nomadic settlers*, a class of people who claim the freedom to move, work, own property and inhabit any place in the world. Despite a superficial openness to new cultures, places and experiences, this tourism appeals to a deep sense of entitlement, where cultural difference, novel experiences and moral validation are commodities to be discovered, collected and worn as trophies of conquest on *Instagram*.

Our itinerary was designed to foster deep, ethical relationships between the delegation and the peoples and places visited based on an appreciation of each person's differentially situated kuleana. Cooper's reaction epitomises the nomadic settler's refusal to be bound by such inconvenient responsibilities. Instead, she seeks the unrestricted freedom to move, to see, to listen, to enjoy and to consume any persons or places she pleases, because, after all, Hawai'i is *for her*.

In the end, the Hawai'i Reality Tour never happened. It seems that it is just too easy to be a tourist in Hawai'i without having to be accountable or coming to terms with their kuleana. The selfish extractive heart of tourism in Hawai'i was summed up by one now infamous *Tik Tok* influencer, who posted a video of herself on a "coronavacation", maskless, drink in hand, taunting, "Go ahead, quarantine me at the Grand Wailea on Maui! I'm gonna live it up! Coronavirus will never win!" (NaturalKaos, 2020).

Conclusion

Hawai'i DeTours are not about tourism. Instead, I would like to consider the Hawaiian word *huaka'i* as an alternative mode of travel for building solidarity. In their groundbreaking anthology *Detours: A Decolonial Guide to Hawai'i*, editors Hōkūlani Aikau and Vernadette Vicuña Gonzalez write:

> A huaka'i is not an empty itinerary or a list of must-dos, but rather a journey defined by intention. A huaka'i is not meant to be an easy walk in the park or a leisurely stroll along the beach. It is demanding. It demands that your journey be deliberate and purposeful, and that you remain open to what you might learn about a place and yourself.
>
> (Aikau & Gonzalez, 2019)

Thus, a huaka'i is more in line with the solidarity delegations that originally inspired the Hawai'i DeTours. Participating in a huaka'i is not about "virtue signalling" or "politically correct" consumption. It requires being responsible and accountable to the people and places on your journey to continue to act in solidarity. Huaka'i also offers a different framework for thinking about the structure and practices of travel. Rather than continuing to market "tourism" as a recreational and consumer-oriented activity, travel should be organised to purposefully make connections and strengthen our mutual responsibilities in an interconnected world.

So, can Hawai'i DeTours unsettle militarisation in Hawai'i? The answer is equivocally, "maybe". As long as Hawai'i remains a lost geography, where settler colonial microaggressions flicker on momentarily, then vanish like a ghost, then, no, the militourism apparatus will gaslight any social justice-oriented tourism initiative, as Cooper has done. However, when combined with other modes of kū'ē and kūkulu, resistance and building, then, yes, the DeTours-as-huaka'i can be an effective practice of political education and movement building. Key elements would include a commitment to the process of building pilina between peoples and places; a creative use of tension against the edges of power to reveal its contradictions; and an openness to contingency and emergent effects. And at its base would be the expectation that anyone privileged enough to experience Hawai'i has responsibilities to stand in solidarity with the land and its peoples.

Note

1 In this paper, I use the Modern Hawaiian Orthography, unless quoting from a source which omitted Hawaiian orthography. In Modern Hawaiian Orthography, the spelling of "Hawai'i" includes the 'okina or glottal stop between the two "i's". However, some proper names may exclude the 'okina.

References

Aikau, H. K., & Vicuña Gonzalez, V. (2019). *Detours: A decolonial guide to Hawai'i.* Durham, NC: Duke University Press.

'Āina Aloha Economic Futures. Retrieved 18 December 2020, from https://www.ainaalohafutures.com/.

Cooper, J. (2011, 12 May). Tropical tour uncovers conflict in Hawaii. *San Francisco Chronicle (SFGate).* Retrieved 18 December 2020 from https://www.sfgate.com/travel/article/Tropical-tour-uncovers-conflict-in-Hawaii-2360780.php

Davis, S. (2015). *The empires' edge: Militarization, resistance, and transcending hegemony in the pacific.* Athens: University of Georgia Press.

Davis, S. (2017). Sharing the struggle: Constructing transnational solidarity in global social movements. *Space & Polity, 21*(2), 158–72. https://doi.org/10.1080/13562576.2017.1324255

Debord, G., & Wolman, G. J. (1956). A user's guide to Détournement. In K. Knabb (Ed.), *Situationist international anthology* (pp. 14–20). Berkeley, CA: Bureau of Public Secrets. Retrieved 18 December 2020 from http://www.bopsecrets.org/SI/detourn.htm

Feeney, P. C. (2009). Aloha and allegiance: Imagining America's paradise. Dissertation, PhD in American Studies, University of Hawai'i at Mānoa.

Ferguson, K. E., & Turnbull, P. (1999). *Oh, say, can you see? The semiotics of the military in Hawai'i.* Minneapolis: University of Minnesota Press.

Foucault, M. (1979). *Discipline and punish: The birth of the prison.* New York: Vintage.

Gonzalez, V. V. (2013). *Securing paradise: Tourism and militarism in Hawai'i and the Philippines. Next wave.* Durham, NC: Duke University Press.

Hau'ofa, E. (1994). Our sea of islands. *The Contemporary Pacific, 6*(1), 148–61.

Hawai‘i State Commission on the Status of Women (2020). Building bridges, not walking on backs: A feminist economic recovery plan for COVID-19. Hawai‘i State Commission on the Status of Women. Retrieved 18 December 2020 from https://humanservices.hawaii.gov/wp-content/uploads/2020/04/4.13.20-Final-Cover-D2-Feminist-Economic-Recovery-D1.pdf.

Imada, A. (2004). Hawaiians on tour: Hula circuits through the American empire. *American Quarterly, 56*(1), 111–49. https://doi.org/10.1353/aq.2004.0009

Imada, A. (2012). *Aloha America: Hula circuits through the U.S. empire.* Durham, NC: Duke University Press.

Kajihiro, K. (2011, 8 December). I Ka Wā Ma Mua, Ka Wā Ma hope: Exploring Pearl Harbor's present pasts. *Hawaii Independent.* Retrieved 18 December 2020, from https://thehawaiiindependent.com/story/i-ka-waa-ma-mua-ka-waa-ma-hope-exploring-pearl-harbors-present-pasts.

Kajihiro, K., & Keko‘olani-Raymond, T. (2019). The Hawai‘i detour project: Demilitarizing sites and sights on O‘ahu. In H. K. Aikau & V. V. Gonzalez (Eds.), *Detours: A decolonial guide to Hawai‘i* (pp. 249–60). Durham, NC: Duke University Press. https://doi.org/10.1215/9781478007203-031

Kinzer, S. (2006). *Overthrow: America's century of regime change from Hawaii to Iraq.* New York: Times Books/Henry Holt.

"Kue: The Hui Aloha Aina Anti-Annexation Petitions, 1897–1898" (1897). Retrieved 18 December 2020, from http://libweb.hawaii.edu/digicoll/annexation/petition.php.

Louis, R. P. (2017). *Kanaka Hawai‘i cartographies: Hula, navigation, and oratory.* Corvallis: Oregon State University Press.

Mak, J. (2015). Creating 'paradise of the pacific': How tourism began in Hawaii. Working paper No. 2015-1. Economic Research Organization at the University of Hawai‘i. Retrieved 18 December 2020, from https://uhero.hawaii.edu/wp-content/uploads/2019/08/WP_2015-1.pdf.

Massey, D. (1991). A global sense of place. *Marxism Today*, June, 24–29.

Mzezewa, T. (2020). Hawaii is a paradise, but whose? *The New York Times*, February 4, 2020, sec. Travel. Retrieved 18 December 2020 from, https://www.nytimes.com/2020/02/04/travel/hawaii-tourism-protests.html.

Mei-Singh, L., & Gonzalez, V. V. (2017). DeTours: Mapping decolonial genealogies in Hawai'i. *Critical Ethnic Studies, 3*(2), 173. https://doi.org/10.5749/jcritethnstud.3.2.0173

NaturalKaos. Tik Tok Post. March 12, 2020. Retrieved 18 December 2020, from https://www.tiktok.com/@naturalkaos/video/6803480232725581061?sender_device=pc&sender_web_id=6880047360928450054&is_from_webapp=v2.

Neason, A. (2020). Greetings from Hawai'i. *Columbia Journalism Review*, Spring 2020. Retrieved 18 December 2020, from https://www.cjr.org/special_report/greetings_from_hawaii.php/

Oliveira, K. R. K. N. (2014). *Ancestral places: Understanding kanaka geographies.* Corvallis: Oregon State University Press.

Osorio, J. H. (2018). (Re)membering ‘upena of intimacies: A Kanaka Maoli Mo‘olelo beyond queer theory. Dissertation. Honolulu: University of Hawai'i at Mānoa.

Pearl Harbor Historic Sites. http://www.pearlharborhistoricsites.org/pearl-harbor/arizona-memorial.

Schofield, M. J. (1873, 15 February). John M. Schofield to my dear general [William T. Sherman]. Box 78. Schofield Papers.

Skwiot, C. (2010). *The purposes of paradise: U.S. tourism and empire in Cuba and Hawai'i*. Philadelphia: University of Pennsylvania Press.

Smith, N. (2004). *American empire: Roosevelt's geographer and the prelude to globalization*. Berkeley: University of California Press.

Teaiwa, T. K. (1999). Reading Paul Gaugin's Noa Noa with Epeli Hau'ofa's kisses in the nederends: Militourism, feminism, and the 'Polynesian' body. In V. Hereniko & R. Wilson (Eds.), *Inside out: Literature, cultural politics, and identity in the new pacific* (pp. 249–64). Lanham, MD and Oxford: Rowman & Littlefield.

Thompson, L. (2010). *Imperial archipelago: Representation and rule in the insular territories under U.S. dominion after 1898*. Honolulu: University of Hawai`i Press.

Trask, H.-K. (1993). Lovely Hula hands': Corporate tourism and the prostitution of Hawaiian culture. In H.-K. Trask (Ed.), *From a native daughter: Colonialism and sovereignty in Hawai'i* (179–97). Monroe: Common Courage Press.

USINDOPACOM Area of Responsibility (2020, 20 October). U.S. indo-pacific command. Retrieved 18 December 2020, from https://www.pacom.mil/About-USINDOPACOM/USPACOM-Area-of-Responsibility/.

Vine, D. (2015). *Base nation: How U.S. military bases abroad harm America and the world*. Kindle. New York: Metropolitan Books, Henry Holt and Company.

Žižek, S. (2006, 9 October). The parallax view: Karatani's "Transcritique. On Kant and Marx". Retrieved 18 December 2020, from http://libcom.org/library/the-parallax-view-karatani-s-transcritique-on-kant-and-marx-zizek.

Section III

Socialising tourism to build better collective futures

9 Public tourism

New forms of tourism after the Great East Japan Earthquake

Shinji Yamashita

Introduction

This year marks the tenth anniversary of the Great East Japan Earthquake. On 11 March 2011, a magnitude 9.0 earthquake struck the Pacific coast of Eastern Japan, followed by towering tsunamis and the meltdown of nuclear reactors in Fukushima. The disaster caused approximately 20,000 deaths and missing people. This was a disaster of unprecedented complexity. The damage was initially estimated at ¥17 trillion, but has so far cost ¥ 32 trillion (Zaimu-shō, 2019). It takes a long time to reconstruct devastated communities; nine years after the disaster, we are still in the process of reconstruction. As of November 2020, there were still about 43,000 evacuees and displaced people (Fukkō-chō, 2020). Disaster-related displacement particularly impacted Fukushima Prefecture, where there are still about 37,000 evacuees and displaced people, including 30,000 people who are forced to stay outside the prefecture (Fukushima Revitalization Station, 2020). It is said that it will take 30 years or more to complete the decommissioning project of the damaged nuclear reactors.

This chapter examines the recovery process from the Great East Japan Earthquake from the viewpoint of tourism.[1] This earthquake had a serious impact on tourism in Japan, with the number of visitors decreasing precipitously in the aftermath. For example, the number of inbound tourists to Japan decreased by over six million people, a 27.8% drop compared to the previous year (JNTO, 2020). The main reason for this decline was that the perception of Japan's safety was shaken, especially because of the meltdown of the nuclear reactors in Fukushima. The tourism industry revealed its vulnerability to the disaster. The disaster, however, brought about various new forms of tourism as well.

Examining the newly emerged forms of tourism after the 3/11 disaster, this chapter aims to demonstrate that tourism could play a positive social role in the recovery process from the disaster. It also contributes to the understanding and solution of public issues caused by the disaster, while redefining the concept of tourism. I refer to this type of tourism as "public

DOI: 10.4324/9781003164616-9

tourism". The chapter demonstrates the important role public tourism can play in encouraging a more social and self-reflexive tourist practice in what German sociologist Ulrich Beck (1992) has called a "risk society".

Volunteer tourism

A month following the disaster, in April 2011, the Japan Tourism Agency (JTA) introduced the *Ganbarō Nippon* (Don't Give Up Japan) campaign to promote tourism in devastated areas as a way to support reconstruction. In June, the National Committee for Reconstruction Vision of the Great East Japan Earthquake (Higashinihon Daishinsai Fukkō Kōsō Kaigi, 2011) proposed a reconstruction plan. With the title of "Utilisation of local tourism resources and creation of new tourism styles", it referred to tourism as follows:

> The tourism industry produces broad-based economic benefits, and together with agriculture, forestry, and fisheries it is a major industry supporting reconstruction. It is expected that local tourism resources including natural views of the beautiful sea, etc., the rich local food culture, indigenous cultural assets such as festivals and shrines and temples, and brands including national parks and World Heritage sites will be widely utilized to create new tourism styles that are only possible in Tohoku and transmit the "Tohoku" brand to the entire country and the entire world.
>
> (2011, p. 30)

What is new, if not necessarily unique, is first of all "volunteer tourism" to the affected region. Although volunteer activities are not new, volunteer tourism, which incorporates volunteer activities into people's vacations/holidays, has been a newly emerging form of tourism in the United States, Europe and Australia (Wearing, 2001). In Japan, volunteers played an important role in the post-disaster reconstruction process, especially after the Great Hanshin-Awaji Earthquake that hit Kobe and surrounding areas in 1995. At that time, a great number of volunteer activists – about 1.3 million from all over Japan – came to assist people affected by the earthquake.

After the Great East Japan Earthquake, approximately 1.5 million volunteers had visited the devastated areas as of July 2016 to support the affected people (Zenshakyō, 2016). The word "volunteer tour" emerged in this process. It was considered that visiting the devastated areas as a tourist would help contribute economically as well as socially to the reconstruction of the local society. In tourism, people usually seek fun and escape during a period of leisure. Volunteer tourists do their volunteer activities in their leisure time, which gives them a sense of making a contribution to humanitarian causes.

Social tourism: in pursuit of *kizuna* or social ties

In the aftermath of the Great East Japan Earthquake, volunteer tourists intended to assist reconstruct communities affected by the disaster. In so doing, tourism was regarded as connecting devastated areas with the rest of the world in rebuilding a new community. Tourism thus contributes to creating *kizuna* or social ties, evoking the feeling of connectedness and togetherness.

One of the fundamental research frameworks within the anthropology of tourism is the relationship between "hosts and guests" (Smith, 1977). However, in commercialised mass tourism there is very little contact between them. In contrast, volunteer tourists may promote social relationships between hosts and guests through assisting affected communities. The concept of tourism as a social activity actually goes back to the origin of modern tourism. In 1841, Thomas Cook, the father of modern tourism, organised the first chartered train tour from Leicester to Loughborough in England to participate in a meeting of the Temperance League (Honjō, 1996, pp. 8–9). The tour organised by Cook was thus intended to be part of his service to greater social well-being.

However, providing support is not necessarily a simple business. Charles MacJilton (2013), director of a Tokyo-based food bank, has reported on "those who reject support". In his support-providing business in the aftermath of the Great East Japan Earthquake, he often experienced that his offer of food was rejected by people saying "we have enough", although in reality they did not have enough. According to him, this may be due to "Japanese culture" in which people are reluctant to accept a "pure gift", i.e. support with no payment or return act required. They feel embarrassed if support is unilateral.

In the relationship between hosts and guests in the context of tourism, if guests/tourists gaze upon their hosts as if they were gazing upon a tourist attraction, the power relationship between hosts and guests may be asymmetrical. Like other exchange systems, those who give a gaze (tourists) are often in a privileged position in relation to those who are given a gaze (hosts). However, if we take hosts as those who show something to guests, the power relationship could reverse; hosts or those who show may be superior to guests, or those to whom things are shown. At any rate, in volunteer, support-providing tourism for community reconstruction, host/guest relations need to be balanced on an equal footing; otherwise volunteering activities do not function.

Manabitabi or study tourism

Another keyword for post-disaster tourism is *manabi* or "study". Tabimusubi (2011), a small tour company in Sendai City, coined the term *manabitabi* or "study tour"_to help the local community in the reconstruction process. Several local NGOs also attempted to make study tour programmes to the devastated areas for visitors to learn about regional ecology and the fishing industry. Minamisanriku-chō Kankō Kyōkai (Minamisanriku Town Tourism Board) has developed a tour programme in which visitors can learn about the disaster through *kataribe* (storyteller) guides. As of October 2020, over 80,000 people have joined since the programme started in 2011 (Minamisanriku-chō Kankō Kyōkai, 2020).

The Human Security Forum (HSF), an NGO established in April 2011 right after the 3/11 disaster by faculty and students of the Human Security Programme at the University of Tokyo, also organised volunteer tours to support the devastated areas in Miyagi Prefecture.[2]_At the early stage of the reconstruction process, from April to November 2011, tours were conducted as "Weekend Volunteer Tours" for the purpose of cleaning up rubble and removing mud from residential areas. Then, in 2012, they were developed into *manabitabi* tour programme. The tour was a three-day event which combined volunteer activities to support holding the summer festival by tsunami refugees from Minamisanriku Town at a temporary housing complex in Tome City. Participants in the programme were mostly university students from Tokyo.

The study tours by HSF were organised in 2013 and 2014 as well. In 2015, however, the *manabitabi* programme in Miyagi came to end, because the summer festival was not held any more due to the fact that many tsunami refugees had left the temporary housing complex at Tome City and moved to new settlements. Therefore, we could say that the role this kind of tour plays may be transitional in the post-disaster tourism context (Tucker et al., 2017). The type of tourism may vary according to the stages of disaster reconstruction.

HSF organised *manabitabi* to Fukushima as well. In 2012, a tour was conducted to attend a seminar on the problems caused by the nuclear plant accident, to listen to stories of disaster experiences and also to watch the revived traditional horse race festival (*Noumaoi*) in Minamisōma City as a symbol of community recovery. Similar tours took place in 2013. In 2014, in conjunction with a research project on the Fukushima disaster,[3] a tour was organised to visit the areas of contamination from the Hamadōri District through Shinchi Town in the Northeastern part of Fukushima Prefecture.

In the fall of 2015, as part of the HSF's *manabitabi* project, a tour was conducted to visit a group of refugees from Tomioka Town in Kōriyama City. The participants in the tour met a group called "Tomioka-machi 3/11 wo Kataru-kai" (Tomioka Group for Narrating the 3/11 Experiences) to listen to their stories of exodus from the nuclear accident area. The following day, the tour participants visited Tomioka Town, located about 10 kilometres south of the Fukushima Dai'ichi nuclear plant. It had become a ghost town because the people were forced to move out as the town was designated a "compulsory evacuation area" from high radiation contamination. In April 2019, the evacuation order was removed, except for *kikan konnan kuiki*, "an area to which it is difficult for the residents to return" due to the high radiation level. However, most people have not been returning to the town. As of August 2020, only 13.2% of the population had returned (*NHK News Web*, 2020a).

In March 2016, another tour to Fukushima was conducted. This time the participants had a lecture by a local junior college teacher on "Fukushima Studies" (*Fukushima-gaku*). Then, the following morning, they visited a coffee shop run by a refugee couple from Iitate Village, another highly polluted area with radiation. Mr. and Mrs. Ichizawa, the shopkeepers, talked about their refugee life. They mentioned the crisis of the family, falling apart because of the different views each family member held about the future. Often, the husband would want to return to the native village of Iitate, while the wife would want to stay in the new residence in Fukushima City and their children would want to raise the next generation in an even safer place in Tokyo. In the afternoon, the tour participants visited Ten'ei Village, which was attempting to produce radiation-free rice. The villagers were also trying to make a new marketing network of consumers to promote their agricultural products, countering the *fūhyō higai* (harmful rumours) that spread after the disaster.

Interestingly, Fukushima Kankō Fukkō Shien Sentā (Support Centre for Revitalising Tourism in Fukushima) ardently promoted this type of tourism. *Fukushima Fukkō Tsūrizumu Gaidobukku* (*The Fukushima Revitalising Tourism Guidebook*), which was published by the Centre, mentioned various tour programmes designed by local people in devastated areas relating to the theme such as "Narrating Evacuation Stories", "Accepting Disaster Refugees", "Learning about Decontamination and Radiation" and "Sharing of Disaster Experiences" (Fukushima Kankō Fukkō Shien Sentā, 2014). The Centre was closed on 31 March 2020 and has been integrated into "Fukushima Prefecture Hope Tourism" since 1 April 2020. Here again, one would observe that the form of tourism evolves in accordance with the stages of disaster reconstruction.

Visiting disaster remains

Kankō, the Japanese word for tourism, originally meant "seeing light". Tourism, however, not only focuses on the bright side of things, but also on the darker side. In this respect, the treatment of the *shinsai ikō* (disaster remains) may be interesting to observe. According to my own personal interview with the *okami* (owner's wife and manager) of Hotel Taikansō in Matsushima Town, it was her policy to welcome guests as if nothing unusual had happened, with the memories of the disaster completely wiped away. In contrast, the *okami* of Hotel Kan'yō in Minamisanriku Town wanted to keep the memory of the disaster alive for a "*kataribe* bus" (storyteller's bus) programme to show the hotel residents around the devastated spots within the town.

Regarding whether or not to keep the memory of the disaster alive, the attitude of the people affected by the disaster is ambivalent. Some want to keep the memory alive, but others do not. The latter was the case with the Daijūhachi Kyōtokumaru, a ship displaced on land by the tsunami in Kesennuma City. The ship was scrapped and removed in September 2013. However, there is a socio-cultural movement to preserve disaster remains as well. The Bōsai Taisaku Chōsha (Disaster Risk Management Centre) in Minamisanriku Town, where ironically 24 town government employees died by incoming tsunamis, was initially earmarked for demolition by the town council, but the decision was reversed and ended up being preserved after protests from the local people. Minamisanriku Town plans to open the *Shinsai Kinen Kōen* (Disaster Memorial Park) in 2021 with a focus on these disaster remains.

Dark tourism: Fukushima as a heterotopia

In tourism studies, this type of tourism is termed "dark tourism" (Lennon & Foley, 2001; Stone et al., 2018). Although dark tourism is a contested term, my position is that this type of tourism is useful for the learning, thinking and understanding of an unfortunate event as well as lamenting over the victims and sympathising with the survivors at the same time. Akira Ide (2013, p. 145), a pioneering scholar of dark tourism in Japan, regards dark tourism as *itamu tabi* (lament tourism).

In this relation, it is worth pointing out that Chernobyl has become a tourist site, attracting many tourists from all over the world. Using Michel Foucault's (1986) concept of "heterotopia", Philip Stone described Chernobyl as follows:

> In summary, Chernobyl is now an-Other place. It exists alongside ordinary spaces of the everyday, yet it is a place where disaster has been captured and suspended. It is a place of crisis, of deviation, of serious reflection…. Indeed, tourists now ritually consume the place as a site of environmental disaster, failed technology and political collapse. Yet, Chernobyl and its dead zone is a surreal space that reflects the reality of our contemporary world – a world exposed by dark tourism.
>
> (2013, pp. 90–91)

Like Chernobyl, Fukushima is also a place of dark tourism, a heterotopia. After the nuclear accident, the Japanese government categorised the radiation-contaminated area into three zones: Zone One, *kikan konnan kuiki*, "an area to which it is difficult for the residents to return"; Zone Two, *kyojū seigen kuiki*, "an area of restricted residence"; and Zone Three, *hinan shiji kaijo junbi kuiki*, "an area in which the authorities are preparing to lift restrictions". Zone One is a barricaded area that no one can enter without permission. Zones Two and Three are areas the residents can freely enter during the daytime, but they may not stay overnight without special permission. These zones are also examples of heterotopias, governed by other principles than those of the ordinary world while existing alongside ordinary space.

The Fukushima Dai'ichi Nuclear Plant (*Ichi Efu*) is located at an overlapping area of two towns, Okuma and Futaba. The area is designated as Zone One, *kikan konnan kuiki*. However, with permission from Tokyo Electric Power Company (TEPCO), one can conduct a tour in the area. By September 2016, during the five-and-a-half years since the accident, approximately 24,000 people had visited this area. In early September 2016, HSF organised a *manabitabi*. First, the participants in this tour visited Iwaki City to listen to an explanation about the current situation of the nuclear plant from the local head officer of the company which is in charge of decommissioning the damaged reactors. The following day, before visiting *Ichi Efu*, they stopped at the J-Village Centre House at Naraha Town, some 20 kilometres away from *Ichi Efu*, to have a brief explanation on the current situation of *Ichi Efu*. J-Village was originally built as a centre for a Japan national football team, but since the nuclear accident it had been functioning as the frontline base for the recovery project from the Fukushima nuclear disaster. Then, the tour group moved to the TEPCO facility at Okuma Town by bus.

Arriving at the *Nyūtaiiki Kanritō* (Control Centre) of the facility, the participants put on protective equipment – cotton gloves, plastic shoe covers and APD (Alarm Pocket Dosimeter). They were required to wear long-sleeved shirts and long trousers covering exposed skin, but there was no special protective clothing or mask for radiation. Then, they took a bus tour to the damaged nuclear reactor buildings and polluted water tanks. It was just like a huge construction site save for the radiation. During the tour the participants remained on the bus, while hearing explanations by the TEPCO staff. The radiation level inside the bus showed approximately 1.3μSv/h at the bus stop near the Control Centre, but reached 40μSv/h near the damaged nuclear reactor buildings. After the tour, they came back to the Control Centre to take off protective equipment and receive a radiation screening test, and then returned to the J-Village Centre House. The total radiation exposure of the tour was approximately 5μSv, which was equivalent to that of a dental X-ray examination.

Fukushima Dai'ichi Nuclear Plant was once promoted with the catchphrase *Genshiryoku, Akarui Mirai no Enerugi* (Nuclear Power, Energy for a Bright Future). But now, it has turned into a symbol of the modern destruction with the new slogan, *Genshiryoku, Hametsu Mirai no Enerugi* (Nuclear Power, Destructive Energy for the Future). *Ichi Efu* is then not only an "other place" but also a "counter-site", i.e. "a kind of effectively enacted utopia" in the real world as a place "simultaneously represented, contested, and inverted" (Foucault, 1986, p. 22). *Ichi Efu* is thus a destination place from which to see the future. For this new future, Hironori Azuma (2013) has proposed a plan in which *Ichi Efu* could become a major dark tourism destination.

Bosai tourism for disaster risk reduction

Bosai is the Japanese word which means "disaster risk reduction". In the efforts to share the experiences of the Great East Japan Earthquake with the world, a new tourism called "*bosai* tourism" has emerged. The tour programme has been developed by the International Research Institute of Disaster Sciences (IRIDeS), Tohoku University, which was established in 2012 right after the disaster. According to the director of IRIDeS Fumihiko Imamura, *bosai* tourism is to combine tourism with disaster risk reduction education. Tourism is an effective tool for disaster risk reduction because people can actually feel the threat of the disaster. If one visits the devastated areas in Tohoku region and listens to the real stories of experiences from the local people, one could learn important lessons for disaster risk reduction in the future. By promoting this kind of tourism, the IRIDeS contributes to spreading Tohoku experiences of the disaster to the world, while promoting the regional development through tourism at the same time (Imamura, 2017).

In cooperation with IRIDeS, Miyagi Prefecture and Sendai City are promoting "Bosai + Tourism" programme to share Miyagi's lessons of disasters and at the same time to introduce the abundant nature, rich culture and diverse industries of the region to the world (Bosai + Tourism, 2020). The programme includes such topics as "Learning about the Different Ways of Evacuation Taken at the Time of the Great East Japan Earthquakes and Tsunami", "Steps Towards Reconstruction Through Business Activities", "Learning about the Reconstruction and the Disaster Preparedness from Previous Disasters" and "Disaster Prevention and Mitigation Lessons from the Disaster: Southern Miyagi Area". One could say that *bosai* tourism is thus an evolved form of *manabitabi* tourism and *kataribe* tourism, which have been already discussed. Disaster brings many negative impacts on the affected communities, but *bosai* tourism transforms them into positive goals of disaster risk reduction towards the future.

Transformation of tourism in the COVID-19 pandemic

The world is now suffering from the COVID-19 pandemic. Originating in China, it has spread rapidly to every corner of the world in the first half of 2020. As of 4 January 2021, approximately 85 million people have been infected and over 1.8 million people have died in the world. In Japan, about 250,000 people have been infected and 3,600 people have died (*NHK News Web*, 2020b). The pandemic is also having an immense socio-economic impact. The United Nations (2020, p. 1) has declared: "this is much more than health crisis. It is human crisis. The coronavirus disease (COVID-19) is attacking societies at their core."

In these circumstances, the rules of the "new normal" are enforced all over the world. They are based on two fundamental principles of "social distancing" and "staying home". These principles particularly affect tourism, because mobility, the foundation of tourism, is forced to be restricted. Tourism is thus one of the hardest-hit sectors by the outbreak of COVID-19. The UNWTO (2020) reported that international tourist numbers could fall 60%–80% in 2020. In Japan, inbound tourism fell by 99.99% in April 2020. Domestic tourism too is affected greatly, to the extent that the number of bookings through major tourist agencies was reduced by over 90% from April to May in 2020 (*Travel Watch*, 2020).

Intending to promote consumption generated by tourism, the Japanese government started the "Go To Travel" campaign in July 2020 with the budget of ¥ 1.7 trillion. The campaign intends to support the severely weakened tourism industry and tourists themselves to increase tourism mobility by subsidising travel expenses. The policy was, however, criticised because the timing was considered inappropriate from the viewpoint of possible risks in spreading the COVID-19 infection. Actually, tours to and from Tokyo, which would be the biggest consumption resource, were excluded as Tokyo was a high pandemic risk region.[4] The COVID-19 situation is changing day by day, so it may be too early to assess the programme.

Many commentators have pointed out that the world is not able to go back to the situation before COVID-19. It presents a transformative moment of restructuring the global economic order in which tourism is included (Hall et al., 2020; Lew et al., 2020). Eijirō Yamakita, president of Japan Tourist Bureau (JTB), the largest tourism company in Japan, stated that it is difficult for the tourism industry to survive the age of COVID-19, and said: "We have already cast away the idea some time ago that JTB is a travel agency. We see JTB today as a company that creates new opportunities for human interaction" (*NHK News Web*, 2020c). As mentioned above, the international tourism industry has been particularly hit hard by COVID-19. In such a situation, Yoshiharu Hoshino, managing director of Hoshino Resorts (2020), Japan, recommends "micro-tourism". This encourages small domestic tourism within a short distance from home to rediscover the value of local tourism resources by local residents. According to him, this is the way of tourism "co-existing with COVID-19".

Public tourism

So far, I have discussed the developments of new forms of tourism after the Great East Japan Earthquake. It must be noted that they contribute to the understanding and solution of public issues caused by the disaster. I refer to this type of tourism as "public tourism". The term public tourism is an idea which came to mind when I was attempting to establish the public anthropology of the Great East Japan Earthquake (Yamashita, 2014a). The disaster raised methodological, theoretical and practical questions regarding how anthropologists should engage with the disaster over a longer time span and what anthropologists can do in collaborative research projects in the future. Putting anthropology to work in the public sphere, I paid special attention to public anthropology that could contribute to the understanding and solution of contemporary public issues beyond the narrow discipline of anthropology (Borofsky, 2007; Yamashita, 2014b). Public tourism is then a useful device to practice public anthropology, bringing together anthropology with tourism to collaboratively work with communities on public issues caused by natural and industrial disaster.

This kind of tourism is, however, usually organised not by commercial travel agents but NGOs, as was discussed in this chapter. In this context, we should recall the Japanese notion of *atarashii kōkyō* (the new public) which emerged in the 1995 Great Hanshin-Awaji Earthquake when over one million volunteer activists came to help the affected people. It marked the advent of a new age of civic activities in Japan. It is termed the "new public" because it is different from the "(old) public", which in Japan was often used in association with the state. By using the term new public, the emphasis was on the citizen or the people. In this new public domain, there emerged new forms of tourism which could be termed comprehensively "public tourism". In Japan, public tourism is thus a project by civic organisations. Public tourism adds to the concept of socialising tourism aiming to promote "tourism for the public good" (Higgins-Desbiolles, 2020). Public tourism too aims for the promotion of the public good by contributing to the understanding and solution of public issues, especially within areas impacted by disaster. The concept also provides new insights into how Japanese tourism, scholarship and society are already engaging with the idea of socialising tourism through public tourism. By doing so, public tourism adds a new dimension to the conceptual and contextual development of socialising tourism.

Tourism in risk society: reflexive tourism

According to Ulrich Beck (1992), contemporary society is a "risk society". Risk is produced by the development of modern society with its advanced science and technology. One of the most serious risks is exemplified by the nuclear accident that occurred in Chernobyl in 1986, which prompted Beck to write his book on risk society. This was the case with Fukushima as well.

The counterpart of risk society is "reflexive modernisation". Beck stated that reflexive modernisation refers to the creative (self-) destruction of an entire epoch of industrial society. He writes: "This new stage, in which progress can turn into self-destruction, in which one kind of modernization undercuts and changes another, is what I call the stage of reflexive modernization"

(Beck et al., 2005, p. 2). Reflexive modernisation, then, is taken as a process by which one modernity is replaced with another: a new modernity. Risk society is therefore regarded as a path to a new modernity that corresponds to the stage of reflexive modernisation.

If risk is a constant in reflexive modernisation, risk management should become another constant as society moves towards a new modernity. In this framework of risk society and reflexive modernisation, tourism also contends with the risk which modern society has produced. Examining the case of Bali, Indonesia, I have proposed a self-reflexive tourism as a possible solution (Yamashita, 2010). The key concept is "reflexivity" for social and ecological justice. Toxic tourism could be an example of reflexive, politically situated social tourism, because it strives for social and ecological justice from the destruction caused by industrial modernity (see Chapter 2; Bigby & Jim, *Tar Creek Toxic Tourism*). I would like to add new forms of public tourism that emerged after the Great East Japan Earthquake as further examples of reflexive tourism, as they represent an emergent reflexive modernity where tourism and tourists are increasingly tied to risk and risk management processes brought on by natural and industrial disaster.

Conclusion

In this chapter, special attention was paid to new forms of tourism that had emerged following the Great East Japan Earthquake: volunteer tourism to assist the devastated communities, social tourism in pursuit of social ties, *manabitabi* or study tourism to the affected areas, dark tourism to visit disaster remains (particularly the damaged Fukushima Dai'ichi Nuclear Plant) and *bosai* tourism for disaster risk reduction. I have argued for the positive social role of tourism in the reconstruction process after the disaster. I have also discussed the transformation of tourism emerging in the midst of the current COVID-19 pandemic. I referred to these forms of tourism comprehensively as "public tourism", where tourism aims to provide the individual tourist with greater understanding and a capacity to engage with others, support community well-being and build a better civic society and public connections.

Through this discussion on public tourism, I have emphasised the reflexive and social character of emerging forms of post-disaster tourism in Japan. Public tourism is thus a project for the promotion of the public good that can direct us towards a more socially driven, "alternative modernity" within tourism: NGO-led rather than private industry, educative travel over rest and relaxation and collaborative with community interests privileged over economic interest. Public tourism is thus a project for the promotion of the public good. Importantly, researchers too are involved in such public tourism, not only as observers but also as participants in tourism. Public tourism is then a practice through which scholars and citizens could collaborate. As such, this chapter demonstrates the important role public tourism can play in encouraging a more social and reflexive tourist practice for the development of "socialising tourism" in disaster recovery environments.

Notes

1 This chapter is a revised version of my former paper (Yamashita, 2016). In revising the paper, I have updated data and added new discussions.
2 The concept of "human security", as a concept that complements "state security", was originally coined by the United Nations Development Programme (UNDP) in 1991 and has been applied mainly to developing countries. However, the Great East Japan Earthquake demonstrated that the concept could also be applicable to a developed country like Japan.
3 This was a research project supported by a JSPS Grants-in-aid for Scientific Research, "Public Anthropology of Disaster Reconstruction: Practicing Creative Development in Fukushima" led by Professor Yuichi Sekiya of the University of Tokyo from April 2013 to March 2016.
4 The restriction was removed from 1 October 2020, but excluded again from 18 December 2020 due to the resurgence in the COVID-19 patients in Tokyo. Further, the Government has now decided to stop the "Go To Travel" campaign all over Japan from 28 December 2020 to 11 January 2021. The period could be extended, depending on the COVID-19 situation.

References

Azuma, H. (Ed.) (2013). *Fukushima dai'ichi genpatsu kankōchika keikaku* [Tourism plan for Fukushima Dai'ichi nuclear plant]. Tokyo: Genron.

Beck, U. (1992). *Risk society: Towards a new modernity.* London: Sage Publications.

Beck, U., Giddens, A., & Lash, S. (2005). *Reflexive modernization: Politics, tradition and aesthetics in the modern social order.* Cambridge: Polity Press.

Borofsky, R. (2007). Defining public anthropology: A personal perspective. Retrieved 6 March, 2012, from http://www.publicanthropology.org/public-anthropology/pdf

Bosai + Tourism (2020). Retrieved 10 August 2020, from https://bosaikanko.jp.

Foucault, M. (1986). Of other spaces. *Diacritics, 16*(1), 22–27.

Fukkō-chō [Reconstruction Agency] (2020). Zenkoku no hinanshasū [The number of evacuees in Japan]. Retrieved 16 December 2020, from https://www.reconstruction.go.jp/topics/main-cat2/sub-cat2-1/20201127_kouhou1.pdf.

Fukushima Kankō Fukkō Shien Sentā [Support Centre for Revitalizing Tourism in Fukushima] (2014). *Fukushima Fukkō Tsūrizumu Gaidobukku* [Fukushima Revitalizing Tourism Guidebook]. Vol.3. Retrieved 26 November 2016, from http://ふくしま観光復興支援センター.jp/about/

Fukushima Revitalization Station (2020). Fukushima-ken kara kengai eno hinaijōkyō [Evacuees outside of Fukushima Prefecture]. Retrieved 16 December 2012, from https://www.pref.fukushima.lg.jp/uploaded/attachment/415713.pdf

Hall, C. M., Scott, D., & Gössling, S. (2020). Pandemics, transformations and tourism: Be careful what you wish for. *Tourism Geographies, 22*(3), 577–98. https://doi.org/10.1080/14616688.2020.1759131

Higashinihon Daishinsai Fukkō Kōsō Kaigi [National Committee for Reconstruction Vision of the Great East Japan Earthquake] (2011). *Fukkō e no teigen: Hisan nonakano kibō* [Towards reconstruction: "Hope beyond the disaster"]. Retrieved 14 December 2020, from http://www.cas.go.jp/jp/fukkou/pdf/fukkouhenoteigen.pdf

Higgins-Desbiolles, F. (2020). Socialising tourism for social and ecological justice after COVID-19. *Tourism Geographies, 22*(3), 610–23. https://doi.org/10.1080/14616688.2020.1757748

Honjō, H. (1996). *Tōmasu kukku no tabi: Kindai tsūrizumu no tanjō* [Thomas Cook's travel: The birth of modern tourism]. Tokyo: Kodansha.

Hoshino Resorts (2020). Hoshino Rizōto ga teiansuru "maikro tsūrizumu" [Hoshino resorts proposes "micro-tourism"] Retrieved 13 June 2020, from https://www.hoshinoresorts.com/information/release/2020/05/90190.htmlhttps://www.hoshinoresorts.com/information/release/2020/05/90190.html

Ide, A. (2013). Dāku tsūrizumu kara kangaeru [Thinking through dark tourism]. In H. Azuma (Ed.), *Fukushima dai'ichi genpatsu kankōchika keikaku* [Tourism plan for Fukushima Dai'ichi nuclear plant] (pp. 144–57). Tokyo: Genron.

Imamura, F. (2017). Tohoku no chiiki-sousei wo ato-oshisuru 'bōsai tzūrizumu' toha [What is *bosai* tourism to promote the reconstruction of Tokoku region?]. *Asu-no-Tsubasa, 9*, 12–14.

JNTO [Japan National Tourism Organization] (2020). Visitor arrivals, Japanese overseas travelers. Retrieved 14 December 2020, from https://www.jnto.go.jp/jpn/statistics/marketingdata_outbound.pdf

Lennon, J., & Foley, M. (2001). *Dark tourism: The attraction of death and disaster.* Stamford, CT: Cengage Learning.

Lew, A. et al. (2020). Visions of travel and tourism after the global COVID-19 transformation of 2020. *Tourism Geographies, 22*(3), 455–66.

MacJilton, C. (2013). Shien wo kobamu hitobito: Hisaichi shien no shōheki to bunkateki haikei [Those who reject support: Obstacles to assist affected areas and their cultural background]. In T. Gill, T., B. Steger, & D. H. Slater (Eds.), *Higashinihon daishinsai no jinruigaku: Tsunami, genpatsu jiko to hisaisha no "sonogo"* [The anthropology of the Great East Japan earthquake: The tsunami, the nuclear accident, and what happened afterwards to the survivors] (pp. 31–62). Kyoto: Jinbun Shoin.

Minamisanriku-chō Kankō Kyōkai [Minamisanriku Town Tourism Board] (2020). Kataribe niyoru manabi no purogramu [Study programme by storytellers of disaster]. Retrieved 26 October 2020, from https://www.m-kankou.jp/tour/storyteller/

NHK News Web (2020a). Retrieved 11 March 2020, from https://www3.nhk.or.jp/news/html/20200311/k10012320891000.html

NHK New Web (2020b). Retrieved 4 January 2021, from https://www3.nhk.or.jp/news/special/coronavirus/data-all/

NHK News Web (2020c). Retrieved 07 August 2020, from https://www3.nhk.or.jp/news/html/20200807/k10012556311000.html

Smith, V. L. (Ed.). (1977). *Hosts and guests: The anthropology of tourism* (1st ed.). Philadelphia: University of Pennsylvania Press.

Stone, P. R. (2013). Dark tourism, heterotopias and post-apocalyptic places: The case of Chernobyl. In L. White & E. Frew (Eds.), *Dark tourism and place Identity: Managing and interpreting dark places* (pp. 79–93). London: Routledge.

Stone, P. R. et al. (Eds.) (2018). *The Palgrave handbook of dark tourism studies.* London: Palgrave Macmillan.

Tabimusubi (2011). Chiiki de manabu, jibun de manabu: Deai to taiken de atarashii jibun hakken no tabi [Learning in local communities, learning by yourself: Discover yourself through travel experiences]. Retrieved 26 November 2016, from http://www.tabimusubi.co.jp

Travel Watch (2020). Retrieved 16 June 2020, from https://travel.watch.impress.co.jp/docs/news/1259314.html

Tucker, H., Shelton, E. J., & Bae, H. (2017). Post-disaster tourism: Towards a tourism of transition. *Tourist Studies*, *17*(3), 306–27.

United Nations (2020). Shared responsibility, global solidarity: Responding to the socio-economic impacts of COVID-19. Retrieved 30 June 2020, from https://www.un.org/sites/un2.un.org/files/un_comprehensive_response_to_covid-19_june_2020.pdf

UNTWO (2020). International tourist numbers could fall 60–80% in 2020. Retrieved 7 May 2020, from https://www.unwto.org/news/covid-19-international-tourist-numbers-could-fall-60-80-in-2020

Wearing, S. (2001). *Volunteer tourism: Experiences that make a difference*. Wallingford: CABI Publishing.

Yamashita, S. (2010). A 20-20 vision of tourism research in Bali: Towards reflexive tourism studies. In D. G. Pearce & R. W. Butler (Eds.), *Tourism research: A 20-20 vision* (pp. 161–73). Oxford: Goodfellow Publishing.

Yamashita, S. (2014a). Introduction. (In special issue: Practicing a public anthropology of the East Japan disaster). *Japanese Review of Cultural Anthropology, 15*, 105–13.

Yamashita, S. (2014b). *Kōkyō jinruigaku* [Public anthropology]. Tokyo: Tokyo Daigaku Shuppankai.

Yamashita, S. (2016). Disaster and tourism: Emerging forms of tourism in the aftermath of the Great East Japan Earthquake. *Asian Journal of Tourism Research, 1*(2), 37–62.

Zaimu-shō (Ministry of Finance, Japan) (2019). Reiwa ninendo yosanan nitsuite [On the budget plan for the fiscal year 2020]. Retrieved 16 November 2020, from https://www.mof.go.jp/budget/budger_workflow/budget/fy2020/seifuan2019/06.pdf

Zenshakyō [Japan National Council of Social Welfare] (2016). Hisaichishien and saigai borantia jōhō [Supporting disaster affected areas and information on volunteer activities]. Retrieved 16 November 2016, from http://www.saigaivc.com/ボランティア活動者数の推移/

10 In search of light

Ecohumanities, tourism and Fukushima's post-disaster resurgence

Adam Doering and Kumi Kato

The search for light in post-disaster Fukushima

The magnitude nine earthquake that hit the Tohoku Region of Japan at 14:46 on 11 March 2011 (3.11) triggered a 38-metre high tsunami and the meltdown of the Fukushima Daiichi Nuclear Power Plant (FDNPP). Coastal villages disappeared, while rural areas surrounding the FDNPP and extending to Fukushima's northwest in the Nakadori Central Region were devastated by nuclear fallout.[1] Dark skies loomed over the horizon in the days following the disaster as radionuclide plumes travelled with the wind, rain and snow over 70 kilometres from the FDNPP, to find landfall on the forests and villages of Nakadori. Throughout the prefecture, the initial phase of recovery consisted of *josen* (decontamination) programmes aimed at collecting and removing five centimetres of contaminated topsoil around residential areas. Along the Hamadori coastline, central and local governments poured considerable resources and funding into the construction of sea dikes, seawalls and embankments.

The Act on Development of Areas Resilient to Tsunami Disasters implemented in December 2011 was designed "to prevent/reduce tsunami disasters in the future, develop a standard institutional system to be utilized nationally and promote 'tsunami resilient city' through 'integrated prevention' incorporating structural and non-structural measures" (Ministry of Land, Infrastructure, Transport and Tourism, 2015, p. 30). Ranging from 7 to 15 metres high, part of this plan includes the construction of 440 discrete seawalls scattered along 405 kilometres of coastline at an estimated cost of over US$12 billion (Kagawa, 2017; Lim, 2018; McNeill, 2016; Yamashita, 2020). These massive sea walls, many still under construction, have been described as providing a sense of security for some (Shimizu, 2018), but for others they have turned seaside towns grey with concrete (Bird, 2013), isolating people and weakening community connections (Koshimura & Shuto, 2015) and making the sea literally and figuratively invisible to these once thriving ocean cultures and communities (Kimura, 2016; Littlejohn, 2020). As Kagawa describes, this top-down view of development "eschews listening

DOI: 10.4324/9781003164616-10

to the voices of local people and fails to appreciate the value of natural environment that made each coastal village and town unique" (2017, n.p.).

Against this background of disaster, loss and managerial recovery, we write in search of light. In its simplest form, the Japanese character 光 (*hikari*) is a noun for light. Like the sun's brilliance that brightens our days to the subtle glow of a firefly that grabs our attention, 光 (*hikari*) – light – brings forth and illuminates. The character is also included in the compound 観光 (*kanko*), meaning tourism or more literally "to see the light". In this chapter, we approach our analysis of tourism and life in post-disaster Fukushima in search of new light. The search for light, however, does not diminish our recognition of many serious political, economic, social and environmental issues that remain unresolved. We acknowledge that the sense of loss is likely to always linger. We are reminded of this in Miri Yu's (2019, p. 53) critically acclaimed novel *Tokyo Ueno Station*, where the Fukushima-born protagonist explains: "Light does not illuminate. It only looks for things to illuminate. And I have never been found by the light. I would always be in the darkness—". Paradoxically, by bringing this to our attention Miri's novel sheds light on such invisibility and darkness.

In a similar way, by targeting issues of inequality, structural violence, alienation and exclusion, the concept of socialising tourism helps us to sense the invisible connections, vulnerabilities and potentialities that are necessary to address if we hope to develop more socially and ecologically just futures (Higgins-Desbiolles, 2020). In addition to positioning people and communities at the centre of tourism decision-making, we argue similar attention could be paid to the often invisible, fragile and yet foundational relations between people and their lands and seas. In this chapter, we situate human-environment relations at the centre of our analysis to shine light on the creativity of people and communities whose care and action animates the ongoingness of life as they reconnect with their natural environments. By doing so, we offer an affirmative, creative and exploratory ethos and methodology for scholars and practitioners of socialising tourism to consider, specifically drawing attention to the importance of an affirmative approach for people and communities in post-disaster tourism settings. We do this by exploring how the search for new light in Fukushima illuminates possible alternative futures, moments of hope and flickers of beauty and creativity, even in the midst of devastation and destruction.

In this chapter we ask: How do mountain and coastal communities in Fukushima reconnect with contaminated lands and seas? How do communities forge new ways of living and belonging with, amongst and against devastated places and spaces? And what role might tourism play in illuminating the emergence of new life in the presence of ongoing disaster? To address these questions, we share stories of life and recovery from Fukushima communities who are reconnecting with the land and sea – spiritually, aesthetically and creatively – and explore the different ways tourism is helping to establish a renewed sense of connection in post-disaster land/seascapes. To give texture to our discussion, examples are drawn from two locations: (1) the creative and artistic tourism undertaken in central Nakadori Region around Iitate Village, which has been an ongoing project of the second author since December 2012; and (2) the new surf tourism developments along the east coast Hamadori Region, where the first author has been engaged in ongoing research with the surfing community at Kitaizumi Beach in Minamisoma City from 2016.

In search of light: the experimental and exploratory ethos of the ecohumanities

The issues and challenges confronting the world today are vast. As Tsing et al. (2017) describe it, we are surrounded by "ghosts" and "monsters" – social division, colonial histories, extreme poverty, industrial farming, anthropogenic climate change, global viruses and radioactive nuclear fallout. Faced with such overwhelming and sweeping devastation, Swanson et al. (2017, p. M2) raise a pertinent question of our times, "In the indeterminate conditions of environmental damage, nature is suddenly unfamiliar again. How shall we find our way"? How *shall* we find our way? How does one engage with the complexities of Fukushima, in all its loss and hope for a better future?

These questions loomed over us as we continued to travel to Fukushima, each time experiencing new layers of complexity. Talking with communities and people about their relationships with the environment over the past decade, one develops a deep sense of what Haraway (2016, p. 132) refers to as "ongoingness", characterised as "somehow cobbling together ways for living and dying well with each other in the tissues of the earth whose very habitability is threatened". Yet in the midst of this felt sense of post-disaster ongoingness arises a sense of urgency and desire for control. Under such duress and uncertainty, it makes sense that a managerial approach tends to dominate the discourses and practices of post-disaster environments: first, as oversimplified *perspectivalism* – is Fukushima polluted or not? (Burch, 2018); second, as *scientification* – precisely how polluted is Fukushima? (Kimura, 2016; Yamaguchi, 2016); and third, as *utilitarianism* – is the imagined future safe yet? (Nancy, 2015). There is always a leaning towards managerialism when disaster strikes as societies seek a sense of being-in-control. This is a reasonable need in the face of such devastation, but may also undermine the necessity of other relations: the socio-technological, embodied, emotional and spiritual relations through which new becomings and worlds emerge out of Fukushima's devastated land and seascapes (Evers, 2019; Kato, 2018a, 2018b; Lin et al., 2018; Littlejohn, 2020; Martini & Minca, 2021). There is a need for new ways of thinking and producing knowledge, different modes of response-ability and experiments with living differently in the world (Gibson et al., 2015; Haraway, 2016).

Listening and sharing stories of people's thoughts, feelings and connections to their lands, seas, histories and traditions, we started to see "recovery" of beauty, hope and life through artistic expression and in relationships with land and sea. These stories were not solutions to problems, but signs of creativity illuminating everyday life realities and futures. In search of light is not a metaphor but a figure, which proved invaluable for helping us to make sense of these experiences and see the world anew. Inspired by feminist more-than-human thinking of Donna Haraway (2016) and Anna Tsing (Tsing, 2015; Tsing et al., 2017) and guided by the experimental and exploratory ethos of the ecohumanities (Gibson et al., 2015; Kato, 2019; Rose, 2010, 2013, 2017), we understand the figure of light as ethos and methodology for socialising tourism to engage with life and the living world in a creative and expansive way. Light has proven to be an important healing figure of our language in these troubling times. The latest example being Amanda Gorman's (2021) poem during the 59th US Presidential Inauguration that captivated the world with her concluding reflection, "For there is always light, if only we're brave enough to see it. If only we're brave enough to be it."

Picture the 15-metre seawall (Figure 10.1) and sense its supposedly reassuring safety and control. In the wake of disaster, "many of us are tempted to address the trouble in terms of making an imagined future safe, of stopping something from happening that looms in the future" (Haraway, 2016, p. 10).

Figure 10.1 Infrastructures of fear. The seven-metre embankment at Toyoma Beach, Iwaki. Photo by Simon Wearne.

Instead of constructing walls, the ecohumanities' "[...] collective inclination has been to go on in an experimental and exploratory mode, in which we refuse to foreclose on options or to jump too quickly to 'solutions'" (Gibson et al., 2015, p. vii). Learning how to dwell a little longer in the paradox of being caught between loss and potentiality, risk aversion and irreconcilable vulnerability may be more conducive to the kinds of thought and action needed to build more liveable futures together.

The search for light involves an affirmative philosophy and ethos for sensing invisible connections and striving towards life and the living world in an exploratory and experimental way (Gibson et al., 2015, p. ii). In her writing on the Northern Australian Yolngu term *bir'yun* (meaning brilliance or shimmering), Rose (2017, p. G53) advocates for a similar approach: "Brilliance actually grabs you. Brilliance allows you, or brings you, into the experience of a vibrant and vibrating world". The earth shimmers, its brilliance calling us into action grounded in care for the land, sea and history. Nothing could be more grounded. The search for light we are striving for here is a similar attentiveness and care for the places we live and travel through and is grounded in relations between humans and their environments. Rose further highlights the ethics of place woven within this line of thought when she writes:

> Rather than hoping that somehow we'll make it through, their way of thinking of the future impels us to take care of the ground right now, right where we are, because we are here...The future is not a promised land waiting for us to arrive, nor does it bear down on us. The future is in the ground. It is life, and it wants to come forth and flourish. The future is creation in everyday life, and like all everyday miracles, it is fragile as it is resilient.
>
> (2010, p. 157)

Instead of seeking out solutions, restorations or reconciliations, the search for light attunes us to the idea of "partial recuperations" (Gibson et al., 2015; Haraway, 2016; Tsing 2015). This requires close attention to what is happening all around us: the small things, moments and fleeting encounters all matter and are deserving of our time and care. The ecohumanities ethos is modest and hopeful, in that it illuminates the historical traces and relations as they unfold in subtle rather than radical ways. The search for light takes notice of the fact that in the midst of terrible disaster – and even within, underneath or alongside neoliberal policies of economic recovery – people and communities may find comfort, meaning and alternative futures in something that was not there before, or was there but is now seen anew (Haraway, 2016; Rose, 2017; Tsing, 2015).

For some, this may sound too optimistic (Padwe, 2019). A common critique directed at more-than-human theories and ecohumanities scholarship is that its focus on non-human agency does not directly or explicitly engage enough with the structures of inequality, productions of marginality and politics more broadly (Padwe, 2019). We agree. Our approach to the ecohumanities here is not directly political in the traditional sense of the term. However, this is not an either-or position. Critical engagement, changing discourses, establishing new economic and social relations and the creation of alternative futures require art and the humanities as much as resistance and political activism. Criticality and creativity do not work in the realm of either-or thinking, especially when one considers the creation of the world as emerging from multiplicity, not binaries (Doering & Zhang, 2018). An ecohumanities approach recognises it is also "important to adopt a reparative rather than purely critical stance toward knowing", maintaining it is equally critical to read for difference as much as dominance (Gibson et al., 2015, p. vii). As Castree (2014, p. 234) further notes, this kind of philosophical foundation creates a distinguishable "structure of feeling", making the ecohumanities' contribution unique amongst other modes of environmental inquiry. In this way, the ecohumanities offers a distinguishable ethos and methodology for socialising tourism in post-disaster environments, both within Fukushima and beyond.

Komorebi: sunlight filtering through the forests of Iitate village

夜空を歩こうか	Shall we walk the night sky
そのようにして見上げる	We look up as if we are
およそ五万人の方々が	Nearly 50,000 people
福島から避難をして	Still living
暮らしている	Away from Fukushima
瞬いている	Shimmering
息を味わい	Hearing their breath
ずっと探している	Been searching all this time
沈黙が沈黙する	Silencing the silence
はるか	So far away
遠くの町のどこかで	In a far town, somewhere
あなたも光に　気がつきましたか	Have you also seen the light?
	(Wago Ryoichi, 2016, pp. 58–59)

The premise of the chapter begins with an art project titled "Fukushima: In Search of New Light", which was designed to highlight the efforts made by local residents returning to resettle in Iitate Village by restarting their traditional business on and with their land. Three artists, poet Wago Ryoichi, percussionist Joyce To and eurythmist Jan Baker Finch, were invited to creatively interpret and express the people's will and efforts to re-engage with their places, which included flower farms, dairy farms, a granite quarry and a stone mason and a blacksmith workshop. These are some of the traditional industries in the village, which were disrupted by the evacuation order that lasted between April 2011 and March 2017. During evacuation, some continued their business by relocating to rented land or stopped their work on farms. The majority of the cattle industry (milk, beef) was discontinued and many of the animals were transferred to other farms, but in many cases had to be left behind (Fukumoto, 2020; Itoh, 2018; Nakanishi et al., 2019). New light here means to relive, restoring the tie with their places, rather than starting something new. Their continued and continuing relationship with their land itself is light.

The three artists visited people's working places over two weeks in August 2019. As a result, the essence of the artistic practices is revealed in sound, rhythm, colour, heat, movement and a combination of these. As part of the continuing support for the recovery from the FDNPP accident (Kato, 2013, 2018a, 2018b), the project aimed to understand people's relationship with their land – knowledge, thoughts, history, feelings and ideas engraved in landscape through creative engagement. Light is support for these people's effort to relive with their places after a long period of absence.

Sunflower resurgence

Sunflowers were planted in many parts of Tohoku to heal the land in the recovery process. This resilient plant is said to grow on salt-affected land and produce radiation-free seeds. Although the positive effects of the flower are scientifically unverified, this sun-chasing bright flower brought much hope and positive prospects for many. Planting is also a process to recover the lost fertility taken from the land that went through the "decontamination" process. Generations of land fertility contained ancestors' history, stories and wisdom, which keeps the Akira family staying with their land.[2] "It will be still some years before the land is good enough to grow rice, so I keep planting sunflowers. [In the meantime] tourists enjoy taking photos here", said Mr. Akira, who made a photographing platform on the edge of his large flower field and made sure to water his flowers on the day of our visit (Figure 10.2).

He feels content: "living away for seven years was too long and I want to stay on the family farm". Family is also returning; his grandson had decided to go to local school rather than boarding in the city school. As the eurythmist Jan moved through the sunflower fields, sounds of the wind, birds and nearby spring tuned in with the percussive rhythm that Joyce created:

明けゆく夜	As the sun rises
光の子らが	Children of the light
先を行く	Are moving forward

(Wago Ryoichi, 2015, n.p.)

Figure 10.2 Sunflower resurgence. Iitate Village. Photo by Simon Wearne.

Blacksmith

A blacksmith workshop, being set up by Kazuki and his wife Hikaru in the former kindergarten in Kusano, Iitate, officially opened in September 2019. It is one of the well-anticipated new beginnings for Iitate since the access restriction was lifted as of 1 April 2017. Kazuki is the fourth generation of the business that started as a saw maker in central Fukushima city. While he continues his training under his master in Tsubame Sanjo, he started to look for a space with possibilities for visitor facilities: a cafe, shop and visitor activities. Iitate offered him this abandoned kindergarten building, which was closed after 3.11. Kazuki and Hikaru saw prospects for a new *mono-zukuri* (craftsmanship) and also an opportunity to encourage visitors to the region.

Kazuki's workshop, renovated from the former kindergarten theatre space with a heavy velvet curtain, is full of old machineries he is slowly restoring. His work represents the local wisdom, *madei*, meaning carefully and mindfully creating with both hands. Kazuki comments: "Machines cannot do the work for you. This is *hand craft* work, even if you use these heavy machines." According to local tradition, blacksmithing is a sacred business combining elements of metal, fire, water, wind, wood and the earth. Starting a fire from the sparkle from hitting the iron intensely is a sacred ritual when starting the furnace. During the performance, the kindergarten theatre is orchestrated with the rhythmic sounds of blacksmith and the fire, blending with gongs, bowls and metals. Jan's bright orange costume moves like a flame. The performance has opened new possibilities and meanings for the space, reinvigorating traditional skills and most importantly, acknowledging the young couple committed to igniting a new fire in this village where people are returning after prolonged evacuation.

Yamada Shrine (Yamada-Jinja)

At the Yamada Shrine in Minamisoma City, the two musicians responded spontaneously to the poem read by Wago Ryoichi. As fate would have it, the performance was serendipitously attended by a group of students from Tokyo who stopped by as part of their study tour. Yamada-jinja was built in 1941 in honour of the Yamada family, who cultivated the region into rice fields. The Yamada Shrine was completely taken by the tsunami, along with the 46 residents. "We honour the great works of people cultivating the land. At the same time we must remember that nature cannot be controlled", explained Priest Mori, who is an archaeologist specialised in the Jomon era. He now surveys the sites where contaminated soils are "temporarily stored" in Okuma town, in the area bordering the FDNPP. Yamada Shrine has become a symbol of hope and resilience in the region; so much so that a group of high school students from Kumamoto in Western Japan fundraised, built and donated a small shrine and a Torii gate in 2012 to temporarily replace the one that had been destroyed while the new Shrine was being relocated to higher ground. The students' torii is engraved with small birds and has become the symbol printed in shrine goods sold to visitors, face towels and *omamori* (amulets/good luck charms). That spirit of care is now imbued in the tourist landscape of the shrine. Towards the end of this creative collaboration, a cool mist came rising from the sea, blanketing us in a gentle cloud, easing the hot, exhausted bodies of the performers, priests, us and audience.

When the wind rises, the wolf appears

The last day of our search for light was back at the Yamatsumi Shrine, where the concept for this project originated (Kato, 2018b), linked to the restoration of the shrine's wolf paintings. These paintings inspired Wago (2015) to write *Wolf*, a *kagura* (sacred music) that was first performed in 2015 at the Mirai-no Matsuri. The *kagura* was inspired by the following story told by the priest's wife, who sadly lost her life in the fire that destroyed the shrine in early 2013. There was an elderly woman from the next village who was an avid worshiper and visited the shrine regularly. She however had very weak legs and was only able to walk with her husband's help. The couple had to walk over the mountain path from the other side of Mt Mora-tori where the shrine is located. The woman used to say:

> As we come towards the shrine, the wind rises and I hear the sound of its movement through the dense long grass. I am able to walk by myself from here. It is the wolf coming to walk with us, it carries me although I never see it.

The wolf, once a sacred animal worshiped and revered with awe, has this invisible power, representing the wild deep mountains that humans are not meant to enter or disturb. In folk tradition, if the wolves came into a human settlement, the villagers would not disturb them.

Incidentally, the trinomial name for the Japanese wolf is *Canis lupus hodophilax*, *hodo* meaning path and *philax* meaning guide. It was only with agricultural cultivation and development that humans encroached on the wolf territory. Wolves were made visible and subsequently were considered dangerous to farm animals, aggressive to humans and transmitters of disease by infecting domestic dogs with rabies. The wolf, once revered and respected, was turned into an evil beast, hated and treated with hostility. When the animal meant to be invisible becomes visible – that is the beginning of an end – it is the species' extinction. The wolf statues at Yamatsumi, which uncannily represents its own species' extinction, now gazes into evacuated villages and contaminated fields and forests that still feel the lingering effects of radiation "damage": the physical, psychological, social and economic destruction brought on by the effects of a nuclear power industry. Wago's *kagura* performance ended powerfully, screaming out for the wolves' (*okami*) return and guidance: "Okami, O--kami, O----kami!" Will wolves walk the villagers' home again? Who will return? There is light in this remembering, in the telling and retelling of this story of human-nature interdependency and indeterminate future.

Kirameki no umi: brilliance and the sea in Minamisoma

> Somewhere, out at the edges, the night
> Is turning and the waves of darkness
> Begin to brighten the shore of dawn
> The heavy dark falls back to earth
> And the freed air goes wild with light,
> The heart fills with fresh, bright breath
> And thoughts stir to give birth to colour.
>
> *Matins* (John O'Donohue, 2008, p. 7)

The excerpt above makes one pause to reflect on the ocean in relation to seascapes, light and colour. These words of eternal return, darkness and brilliance are insights gained through a life lived by the sea. The Fukushima surfers' return to the ocean is also a story full of anxiety of the unknown and deep concern about the death that surrounds their enjoyment of the sea – *waves of darkness*. But like the wolves' howl in Iitate, the sea too acts as a call and guiding light for the surfers' return to Fukushima. Two years before the "Fukushima, in search of new light" project began, the first author (henceforth singular pronouns will be used to allow for personal reflection) started exploring the human-ocean connections at Kitaizumi Beach in Minamisoma City. Located 25 kilometres north of the FDNPP, I began listening to the experiences of Fukushima surfers who chose to stay or return to a "contaminated" sea.

Staying with "contaminated" seas

Waves brought on tragedy, but they also open possibilities for partial recovery by calling on people to communicate with and care for the sea. Rumours, toxic uncertainty/realities and loss, "contaminated" the social and physical spaces of surfing in the region for the first few years (Evers, 2019). However, despite the confusion and uncertainty, one thing remained: the sea. Surfers described this bond with the waves and sea as being an important reason that encouraged them to stay in Fukushima. Research on lifestyle sports like surfing argues participants develop unique bonds with the natural environment over time and may even feel a part of their living environments when engaged in the activities (Thorpe & Rinehart, 2010). Thorpe (2014) uses rhythm analysis to further explain how people engaged in these activities are also among the first to return and re-engage with post-disaster ecologies (Thorpe, 2014). During initial interviews and observations at Kitaizumi Beach, this was certainly the case. Surfers filled the parking lots at Kitaizumi Beach, the lively and open atmosphere making it a unique culture amongst Japan's highly localised surf cultures (Doering & Evers, 2019).

Their return, however, was complex. When asked of his return to the surf, Sei, a leading member of the Fukushima Surf Association in his 40s, replied:

> It was complicated…It would be a lie to say I wasn't concerned [about the radiation]. Some of the other surfers entered the sea after a year and a half, or maybe two years. So, when I entered, I was half excited and half anxious.

The Fukushima surfing community decided to follow the practice of *sankaiki*, which literally means the third mourning, a memorial service customarily held for a departed soul in accordance with the tenets of Buddhism. This meant surfers were not supposed to enter the sea for three years; this was a time for healing. While most followed this practice, stories circulated of people surfing as early as one month following the disaster. A respected surfer in the area explained how he would "sneak in the surf before people were awake". There was still a stigma around entering the sea where so much loss had occurred. In contrast to news reports characterising Fukushima surfers as radical risk-takers defying the dangers of radioactive water (Gartside, 2016; Laforgue, 2016), the prevailing concern was more about respect for lost lives than radiation. A leader in the surfing community, who followed the practice of *sankaiki*, put it bluntly: "How could you walk over a graveyard to go surfing"? This was a sentiment shared by other non-surfing members in the community.

Kazu, a first-generation surfer now in his 60s, began surfing at Kitaizumi Beach in 1975. He explained how he has surfed at this beach nearly every day for 43 years. Kazu told me of the difficulties of evacuation and his eventual return to the sea. In the following months of the disaster, he moved between Tokyo, Sendai and other temporary shelters. He had lost everything: his surf shop, house, mother and later on his father. He initially evacuated, only to return a few months later for the summer swell. Kazu felt adrift being away from his wave: "I couldn't stand it. I thought if I didn't start surfing again soon, I wouldn't be able to surf again for the rest of life." He had lost everything; surfing was his last beacon of light.

Searching for the light allows us to see Kazu not only as a statistic – one of who returned – but also reveals the deep human-ocean bonds that called on some surfers to stay and/or return to the sea. Staying with the sea is a practice of partial recovery characterised by a "refusal to deny irreversible destruction, and refusal to disengage from living and dying well in the presents and futures" (Haraway, 2016, p. 86). Surfers returned to the sea without fully knowing the details of radiation and agonising deeply about the loss surrounding their enjoyment of the sea. Politically, surfers explained that they could do little to protest against the plans to build seawalls or to put a stop to nationalised nuclear power. Kazu described that although he does not have a clear idea of what the future may bring, in fact, stating there will be no returning to normal for Fukushima, his ongoing presence in the sea is itself an important action:

> We have to keep doing what we're doing and pass on the goodness of what we're doing to those who are a little younger than us. Then, those who are younger will pass it on to the next generation. And if we do that, I think people will return again someday.

There are many kinds of activism, some requiring precisely these kinds of patient, persistent and partial recuperations that can heal not only people and communities, but the devastated lands and seas. The bodily presence of surfers at Kitaizumi Beach keeps open subtle possibilities for such partial recovery. Without surfers, the bay may have succumbed to the fate of other beaches in the area: fortified, walled and reimagined as a techno-future, solar-panelled landscape or a possible site for storing nuclear waste. The ongoing presence and actions of the surfers resemble Lora-Wainwright's (2017) concept of resigned activism. The term acquiescence resonates with the Fukushima surfers, referring to the reluctant acceptance of something without active/visible protest. Just staying with the sea takes an enormous amount of effort and courage, an act that is itself already a powerful form of engagement and "action". Acquiescence is not the opposite of practical action/resistance; rather, staying with the sea could also be construed as an act of resistance and partial recuperation, one that forms the foundation of an informed, multisensory and nuanced response-ability to living with socially and environmentally "contaminated" environments (Haraway, 2016). Recovery – or more accurately partial recuperation from disaster – depends on these more-than-human relations (Tsing, 2015).

Surf tourism as resurgence: lifesavers and the return to the sea

Let's return to the August 2019 "In Search of New Light" fieldwork. Being able to experience the artistic engagement of the three artists – poet Wago Ryoichi, percussionist Joyce To and eurythmist Jan Baker Finch – with the landscapes and seas had a profound influence on the way the first author started to see and experience the coastline in and around Minamisoma during the search for light project. The experimental and exploratory way the artists engaged with the seascape encouraged me to see familiar places anew. On 9 August 2019, our group woke just before dawn making it to the impressive seawall in Toyoma Beach (Figure 10.1) just before the sun peaked over the ocean horizon. It was my role to take the artists to the sea. Arriving at the coast, our attention was immediately directed towards the domineering presence of the seawall. Jan put on her "sand shoes" and precariously danced along the seawall. Each shoe was weighed down with two kilograms of sand; the scene felt heavy. Like an embodiment of the seawall itself, it is difficult to feel light in the presence of so much out-of-place sand.

Exhausted by investing all our energy in the wall, we turned our attention to the sea (Figure 10.3). Jan dumped the sand out of her shoes and skipped into the Pacific.

The lightness of movement, the colours, the brilliance, all this stood in stark contrast to the dreary wall. There were no surfers in the sea; rare, considering it was mid-summer at one of the most popular beaches in the area: just the wall, sea and lightness of expression. Jan turned and described the scene, exclaiming that despite the wall "the ocean keeps on oceaning". This moment, however fleeting, brought an awareness on how to empathise with this natural/man-made disaster and massive seawall, to acknowledge the incompleteness of the situation and to experience in an embodied sense the brilliance and resilience of the eternally returning sea.

A month earlier, on 20 July 2019, an *umibiraki* (beach opening) ceremony was held to officially open the beach for public use for the first time in eight years. The Minamisoma City Tourism Association (2020) estimated 37,732 people attended the *umibiraki* event, and importantly, it was estimated around 30% of attendees were children. The beach opening event was largely made possible through the hard work of the municipally funded

Figure 10.3 Brilliance and the sea. Toyoma beach, Iwaki, 9 August 2019. Photo by Adam Doering.

"Happy Island Surf Tourism Non-Profit Organisation (NPO; the two characters of Fukushima translate as "Happy Island"). In 2019, Minamisoma City awarded the Happy Island Surf Tourism NPO ¥ 26,000,000 over five years (US$260,000) for surf tourism development. A large percentage of this funding was being allocated to employing the lifesaving team at Kitaizumi Beach. The beach opening was the first day for these lifesavers; safety was paramount and anxiety was high. As the team leader of the lifesavers explained in the evening after the first day: "Can you imagine if someone had died today"?

The objective of pre-disaster surf tourism development in Minamisoma was to revitalize the declining rural economy and diminishing population. Post-disaster surf tourism shifted its focus towards oceanic knowledge, safety and youth education. As Haru described: "Surfers should share their knowledge, not only to 'revitalise' [the economy], but relearn our knowledge of the sea. We have to rethink our lifestyle. Considering Japan is an island country, we have to relearn how to live on island". Surf tourism may not be the answer to all the serious social issues Fukushima continues to face, but it does provide the time, compassion and material support required for these lifesavers to begin rebuilding lives with the sea in hope that locals and tourists would soon return as well.

Intergenerational knowledge sharing, however, requires someone to whom knowledge is passed down. Fukushima surfers repeatedly expressed the concern that as a result of the disaster, they may lose an entire generation of young surfers. So, while the objective was now to create spaces for youth involvement in outdoor ocean-based activities, youth surfers at Kitaizumi Beach were extremely rare. With the exception of the *umibiraki* event, in five years of visiting Kitaizumi, I had never seen a youth in the sea. This makes the unexpected event that follows all the more impactful and surprising.

Dancing with the sea is a form of brilliance, the kind described earlier by Rose (2017, p. G53) that "actually grabs you" bringing one "into the experience of a vibrant and vibrating world". I was still vibrating with these thoughts and emotions when I returned to Minamisoma on 11 August, just two days travelling with the artists in search of light, to visit the lifesavers at Kitaizumi Beach. I arrived at 7:30 in the morning, just before the 8:00 lifesaver start time. On the way to the beach, the lifesavers were already in the water where a small group was gathered at the shore (Figure 10.4).

The young girl about to stand up on the surfboard is nine years old. She is being taught and guided by the Kitaizumi lifesaver team who had agreed to provide her surf lessons prior to the start of their shift. An ethics of care dominates this ocean scene, contrasting starkly with the fear embodied in a seawall. Her mother told us the story of how her daughter was born on 10 March, the day before the triple disaster. Her family was from Minamisoma, but she did not return for three years. It had been nine years since she returned to the sea. Her first visit was during the *umibiraki* beach opening

Figure 10.4 Ethics of care. Youth returning to the surf at Kitaizumi Beach, Minamisoma, 11 August 2019. Photo by Adam Doering.

event and she had decided it would be good for her kids to try surfing. The decision did not appear particularly difficult for her; this day was a normal day; it was as if she could have been taking this surf lesson anywhere. When the young girl stood on the surfboard for the first time, the radiance in the smiles of the lifesavers' faces and cheers from onlookers on the beach spoke volumes. Amongst daily and ongoing worry and concern, there could not be a more fitting way to start their shift dedicated to saving lives and to embody an ethics of care for the sea and for generations to come. At that moment, in that place, where dark, cold waves once again hit the shore, there is light: "The heart fills with fresh, bright breath. And thoughts stir to give birth to colour" (O'Donohue, 2008, p. 7). Like waves, light comes and goes in swells. But in that moment – following years of isolation and discrimination, of uncertainty and anxiety, of difficult life decisions and determination just to simply stay with the waves in Minamisoma – the sea and surfers were brilliance.

Reflections: socialising tourism and the search for light

This chapter offered a curation of stories and performances from post-disaster Fukushima – sunflowers, sand shoes, wolf extinction, surfing children and The Great Wall. After ten years of a persistent flood of media images of the disaster, what does one think of when mentioning "Fukushima"? One aim of our search for light was to help readers feel, think and see Fukushima's devastated lands and seas anew. The art of noticing how human-environment relations persist in the midst of disturbance is a critical first step for opening spaces of possibilities and potential alternative futures (Tsing, 2015).

We attempted here to embody the ecohumanities ethos in our approach to writing and in our engagement with the people, lands and seas of Fukushima. The search for light guided us to look and listen for "the life-giving potentialities...opening ourselves to what can be learned from what is already happening in the world" (Gibson et al., 2015, p. ii). While acknowledging the ongoing necessity of critical analysis for the socialising tourism concept (Higgins-Desbiolles, 2020), the search for light also highlights the importance of creative and reparative methodologies for socialising tourism, particularly in post-disaster settings. It is a methodology characterised by Gibson et al. (2015) as three overlapping commitments: (1) to develop new terms, figures, phrases and language needed to respond to a changing world; (2) to enter into uncertainty and make risky attachments by staying with the trouble (Haraway, 2016); and (3) to join and support concerned others in the plight to rebuild better lives amidst disaster. The ecohumanities offers a distinguishable ethos and methodology for socialising tourism in post-disaster environments and beyond. Writing can be light.

We want to draw attention to two final reflections from this work. First, recovery – healing, partial recuperation, rebuilding, rethinking

– requires a concerted focus on human and more-than-human relations. As we experienced, in many instances it was the human-sea-soil relations that kept people grounded to their places, while also offering hope that one day others, including tourists, will also someday return. Second is the issue of ecohumanities, tourism and critical praxis. Some may understand post-disaster tourism in Fukushima as a form of disaster capitalism (Klein, 2007). Tourism may not be an "answer" or final solution for Fukushima, but that kind of logic misses the point. Such an approach undervalues the labour, time and compassion that the people and communities of Fukushima clearly demonstrated towards their environments in their modest tourism endeavours. Tourism provided surfers with ocean-based employment, affording another possibility to continue their lives with the sea. The possibility of tourism also offered farmers, priests and craftspeople added inspiration to restore their ties with their lands in creative ways rather than moving away or starting something entirely new. These modest engagements with tourism provided the gift of time, compassion and material support to continue living *with* soil and sea. By attending to these often invisible, fragile and yet foundational relations between people and their lands and seas, more ecologically just tourism futures become possible. Tourism can be light.

Notes

1 To see how the restricted zones have changed over time in this region, see Fukushima Prefecture's (n.d.) *Hinan Shiji-to no Ikisatsu* (*Background of Evacuation Orders*). Retrieved 12 January 2021 from https://www.pref.fukushima.lg.jp/uploaded/attachment/254764.pdf.

2 With the exception of published authors and the three artists, all names of people are pseudonyms.

References

Bird, W. (2013). Post-tsunami Japan's push to rebuild coast in concrete. *The Asia-Pacific Journal: Japan Focus, 11*(2). Retrieved 21 October 2020, from https://apjjf.org/2013/11/21/Winifred-Bird/3945/article.html

Burch, K. A. (2018). Eating a nuclear disaster: A vital institutional ethnography of everyday eating in the aftermath of Tokyo Electric Power Company's Fukushima Daiichi Nuclear Power Plant disaster (Thesis, Doctor of Philosophy). University of Otago. Retrieved from http://hdl.handle.net/10523/8084

Castree, N. (2014). The Anthropocene and the environmental humanities: Extending the conversation. *Environmental Humanities, 5*, 233–60.

Doering, A., & Evers, C. (2019). Maintaining masculinities in Japan's transnational surfscapes: Space, place, and gender. *Journal of Sport and Social Issues, 43*(5), 386–406.

Doering, A., & Zhang, J. J. (2018). Critical tourism studies and the world: Sense, praxis, and the politics of creation. *Tourism Analysis, 23*(2), 227–37.

Evers, C. (2019). Polluted leisure and blue spaces: More-than-human concerns in Fukushima. *Journal of Sport and Social Issues*. https://doi.org/10.1177/0193723519884854

Fukumoto, M. (2020). *Low-dose radiation effects on animals and ecosystems. Long-term study on the Fukushima nuclear accident*. Singapore: Springer Nature.

Gartside, L. (2016, 18 October). The fearless surfers of Fukushima. *Wavelength Magazine*. Retrieved 10 January 2021, from https://wavelengthmag.com/the-fearless-surfers-of-fukushima/

Gibson, K., Rose, D. B., & Fincher, R. (Eds.). (2015). *Manifesto for living in the Anthropocene*. Brooklyn: Punctum Books.

Gorman, A. (2021, 20 January). *The hill we climb*. Retrieved 30 January 2021, from https://edition.cnn.com/2021/01/20/politics/amanda-gorman-inaugural-poem-transcript/index.html

Haraway, D. J. (2016). *Staying with the trouble: Making kin in the Chthulucene*. Chicago, IL: Duke University Press.

Higgins-Desbiolles, F. (2020) Socialising tourism for social and ecological justice after COVID-19, *Tourism Geographies, 22*(3), 610–23. https://doi.org/10.1080/14616688.2020.1757748

Itoh, M. (2018). *Animals and the Fukushima nuclear disaster*. London: Palgrave McMillan.

Kagawa, F. (2017). Concrete walls, cherry trees and post-tsunami reconstruction in Japan. *Sustainability Frontiers*. Retrieved 22 March 2018, from http://www.sustainabilityfrontiers.org/index.php?page=view-from-the-linden-barn-6-december-2017

Kato, K. (2013). As Fukushima unfolds: Media meltdown and public empowerment. In L. Lester & B. Hutchins (Eds.), *Environmental conflict and the media* (pp. 201–14). New York: Peter Lang.

Kato, K. (2018a). Debating sustainability in tourism development: Resilience, traditional knowledge and community: A post-disaster perspective. *Tourism Planning & Development, 15*(1), 55–67.

Kato, K. (2018b). Restoring spiritual resilience in Fukushima. In A. Lew & J. Cheer (Eds.), *Tourism and resilience in environmental changes* (pp. 236–49). London: Routledge.

Kato, K. (2019). Gender and sustainability–exploring ways of knowing–an ecohumanities perspective. *Journal of Sustainable Tourism, 27*(7), 939–56.

Kimura, A. H. (2016). *Radiation brain moms and citizen scientists: The gender politics of food contamination after Fukushima*. Durham, NC: Duke University Press.

Klein, N. (2007). *The shock doctrine: The rise of disaster capitalism*. New York: Metropolitan Books.

Koshimura, S., & Shuto, N. (2015). Response to the 2011 Great East Japan earthquake and tsunami disaster. *Philosophical Transactions of the Royal Society A: Mathematical, Physical and Engineering Sciences, 373*(2053), 20140373. https://doi.org/10.1098/rsta.2014.0373

Laforgue, E. (2016, 29 August). Fukushima's surfers riding on radioactive waves. *Al Jazeera*. Retrieved 5 January 2021, from https://www.aljazeera.com/gallery/2016/8/29/fukushimas-surfers-riding-on-radioactive-waves

Lim, M. (2018, 9 March). Seven years after tsunami, Japanese live uneasily with seawalls. *Reuters*. Retrieved 5 May 2018, from https://uk.reuters.com/article/us-japan-disaster-seawalls/sevenyears-after-tsunami-japanese-live-uneasily-with-seawalls-idUKKCN1GL0DK

Lin, Y., Kelemen, M., & Tresidder, R. (2018). Post-disaster tourism: Building resilience through community-led approaches in the aftermath of the 2011 disasters in Japan. *Journal of Sustainable Tourism, 26*(10), 1766–83.

Littlejohn, A. (2020). Dividing worlds: Tsunamis, seawalls, and ontological politics in northeast Japan. *Social Analysis, 64*(1), 24–43.

Lora-Wainwright, A. (2017). *Resigned activism: Living with pollution in rural China.* London: MIT Press.

Martini, A. & Minca, C. (2021). Affective dark tourism encounters: Rikuzentakata after the 2011 Great East Japan Disaster, *Social & Cultural Geography*, 22:1, 33–57, DOI: 10.1080/14649365.2018.1550804.

McNeill, D. (2016, 5 March). Japan's sea wall: Storm brews over plans to construct giant £5bn barrier against tsunamis. *Independent.* Retrieved 10 February 2019, from http://www.independent.co.uk/news/world/asia/japans-sea-wall-storm-brews-over-plans-to-construct-giant-5bn-barrier-against-tsunamis-a6914781.html

Minamisoma City Tourism Association (2020, 16 January). Personal interview [Personal interview].

Ministry of Land, Infrastructure and Transport and Tourism [MLIT] (2015). Recent policy changes regarding tsunami disaster countermeasures. Retrieved 10 March 2018, from https://www.mlit.go.jp/river/basic_info/english/pdf/conf_10-3.pdf

Nakanishi, T., O'Brien, M., & Tanoi, K. (Eds). (2019). *Agricultural implication of Fukushima nuclear accident (III). After 7 years.* Singapore: Springer Nature.

Nancy, J. L. (2015). *After Fukushima: The equivalence of catastrophes.* New York: Fordham University Press.

O'Donohue, J. (2008). *To bless the space between us: A book of blessings.* New York: Doubleday.

Padwe, J. (2019). Book review: The mushroom at the end of the world: On the possibility of life in capitalist ruins, *The Journal of Peasant Studies*, 46(2), 433–7.

Rose, D. B. (2010). So the future can come forth from the ground. In K. D. Moore & M. Nelson (Eds.), *Moral ground: Ethical action for a planet in peril*, (pp. 154–7). San Antonio, TX: Trinity University Press.

Rose, D. B. (2013). Slowly: Writing into the Anthropocene. *TEXT Special Issue 20: Writing Creates Ecology and Ecology Creates Writing, 20*, 1–14.

Rose, D. B. (2017). Shimmer: When all you love is being trashed. In A. L. Tsing et al. (Eds.), *Arts of living on a damaged planet: Ghosts and monsters of the Anthropocene* (pp. G51–63). Minneapolis: University of Minnesota Press.

Shimizu, K. (2018, 28 April). *Shiroi kabe chiiki mimamoru* [The white wall watching over the community]. *Yomiuri Shimbun*, p. 10.

Swanson, H. A., Tsing, A. L., Bubandt, N., & Gan, E. (2017). Introduction: Bodies tumbled into bodies. In A. Tsing, H. Swanson, E. Gan, & N. Bubandt, N. (Eds.), *Arts of living on a damaged planet: Ghosts and monsters of the Anthropocene* (pp. M1–12). Minneapolis: University of Minnesota Press.

Thorpe, H. (2014). *Transnational mobilities in action sport cultures.* London: Palgrave Macmillan.

Thorpe, H., & Rinehart, R. (2010). Alternative sport and affect: Non-representational theory examined. *Sport in Society: Cultures, Commerce, Media, Politics, 13*(7–8), 1268–91.

Tsing, A. L. (2015). *The mushroom at the end of the world: On the possibility of life in capitalist ruins*. Oxford: Princeton University Press.

Tsing, A. L., Bubandt, N., Gan, E., & Swanson, H. A. (2017). *Arts of living on a damaged planet: Ghosts and monsters of the Anthropocene*. Minneapolis: University of Minnesota Press.

Wago, R. (2015). Shiro-hadashi, Mirai no matsuri [Festival for the future]. Retrieved 13 October 2020, from www.mirainomatsuri-fukushima.jp

Wago, R. (2016). Kino-yorimo yasashiku naritai [I want to be more kind than yesterday]. Tokuma Shoten.

Yamaguchi, T. (2016). Scientification and social control: Defining radiation contamination in food and farms. *Science Technology & Society, 21*(1), 66–87.

Yamashita, H. (2020). Living together with seawalls: Risks and reflexive modernization in Japan. *Environmental Sociology, 6*(2), 166–81.

Yu, M. (2019). *Tokyo Ueno station*. London: Tilted Axis Press.

11 Socialising animal-based tourism

Carol Kline

Introduction: animal oppression in tourism

The documentation and categorisations of animal use within tourism are extensive and have been steadily growing since the 1980s. The tourism literature has approached the topic from various moral and ethical lenses (Winter, 2020), cataloguing various forms of animal tourism (e.g. animals as food, working animals, animals in entertainment, wildlife tourism), classifying industry practices and teasing out various levels of animal welfare violations. In general, most "pro animal welfare" writing has attempted to shift our perspective from an anthropocentric to a more biocentric or animal-centric one, advocating for more justice (Sheppard & Fennell, 2019) and less neoliberal approaches of exploitation and commodification (Duffy, 2013; Higgins-Desbiolles, 2020). The recent momentum in animal tourism studies is significant; therefore, combined with the current growing public awareness of climate change, animal welfare and racial justice, the premise of this chapter is that the time is right to go one step further. Just as we must evoke urgency in changing the operations of industries such as mining, technology, manufactured goods, modern agriculture and luxury services – because of their severely flawed supply chains and contribution to climate change and social injustice – we must also swiftly reckon with tourism's continual exploitation of natural resources and fellow global citizens (Büscher & Fletcher, 2017). We must scrap or significantly revamp the old models of management and profiteering – project by project and nation by nation – and consider the socialisation of animal-based tourism.

Indeed, most of the ethical problems of "capitalising on" animals correlates to other abuses, oppressions and injustices (e.g. female exploitation, unfettered violence and death, unequal distribution of wealth, wilful promotion of ignorance, intentional class segregation, environmental hazards) and are rooted in an array of related "ism's": capitalism, neoliberalism, anthroprocentrism and perhaps the bedrock of them all, exceptionalism. From these foundational "ism's" flow many manifestations of injustice: racism, sexism, able-ism, homophobia and speciesism (Adams, 1991; Donovan & Adams, 2007; Büscher & Fletcher, 2017; Roberts, 1997).

DOI: 10.4324/9781003164616-11

Speciesism is the belief that some species are "better than" others; most notably, *homo sapiens* are better than all others. The upshot of this belief is that the interests of non-human species are subjugated to those of humans (Holden, 2019). Under systems of capitalism and neoliberalism, this has turned out to be detrimental for individual animals, for species and for collective ecosystems. The concept of speciesism was coined by Dr. Richard Ryder in 1970 and has since been incorporated into the vernacular of animal advocates and animal studies scholars around the world (Ryder, 2004, 2017). Some common trappings of speciesism include the belief that humans are the only animals who are self-aware and autonomous, have language, have emotions, use tools and can build and create things (*The Superior Human?*, 2012*)*. Speciesism within tourism is embedded within most encounters that a visitor has with an animal and typically the direction of the relationship is one way – from the user to the "resource" being used. To counter a human supremacist worldview is to instil an ethic for species justice as a step towards socialising animal-based tourism.

Species justice

Animal justice is used broadly to denote that non-human animals should have a measure of fair treatment afforded to them by humans. Species justice can similarly mean that all creatures should have access to fairness and equity regardless of their species membership. However, species justice is used most often to promote fair treatment of non-human animals by humans rather than promoting the notion of equity relegated to cats and dogs, for example. For this reason, the terms animal justice and species justice will be used here interchangeably. And while the notion of the animal continuum (the premise that humans do not consider all non-human animals equally but allocate higher status to animals who are charismatic, cute, useful or docile) is real and varies from culture to culture, I will only treat this topic summarily later in the chapter. The concept of justice is relevant to the discussion as its principles are the driving motivator behind socialising tourism, and specifically animal-focused tourism. Further, a political conception of justice is needed to frame animal-based tourism, in that it is wholly an economic, political and cultural phenomenon that cannot be unlinked from moral and ethical agendas of animal welfare and animal rights. Rawls (1987, p. 3) puts it this way: a "political conception of justice is … of course, a moral conception, *(but)* it is *(also)* a moral conception worked out for a specific kind of subject, namely, for political, social and economic institutions". In this chapter, I will address how a political system of species justice could be introduced within the confines of tourism frameworks that we already have. This is with a goal of mitigating the atrocities perpetrated on animals while maintaining fairness and equity for human communities (acknowledging that we should first expand our conception of communities to include animals as well as the ecosystems we all inhabit).

In 2002, Scheyvens introduced the notion of justice tourism as an ethical approach to tourism that encompasses social change through political activism by organisations and by tourists. Jamal (2019, p. 221) contended that "too little attention has been given in tourism literature to ethical and justice theories, philosophy, animal welfare (and rights), diverse world views, climate change and social action". Higgins-Desbiolles (2008) advocated for justice tourism to address the social inequities caused by capitalism and globalisation. The concept of justice *within* tourism has been applied to animals, both in conjunction with and independently from a discussion of ethics. For example, ecological justice and animal ethics were joined within Jamal's (2019) reflection on tourism literature of the past 75 years. Holden (2019, p. 699) likewise purported: "it is necessary to recognise that environmental ethics often cannot be divorced from anthropic issues of social justice, and that the basic interests of both nature and humans have to be accounted for". Fennell and Sheppard (2019) addressed the "justice agenda" of animal advocacy organisations and animal sanctuaries who advocate for animal welfare guidelines and who emphasise ethics of care in practice. They contrast the justice-in-practice activities of these organisations with the codes of conduct devised by global authorities that only purport an instrumental view of animals embedded in tourism.

Most recently, Fennell and Sheppard (2020) introduced a justice framework that may be applied to animal-based experiences in tourism that includes four levels: no justice, shallow, moderate, and deep justice. The framework can serve well as a departure point for academics and industry to assess the fairness afforded to animals within various types of tourism activity. Further, Fennell and his co-authors have a range of macro approaches to how industry contends with animal ethics in tourism (Fennell, 2019, 2020; Garrod & Fennell, 2004; Malloy & Fennell, 1998; Sheppard & Fennell, 2019). This includes analyses of various codes of ethics, codes of conduct and tourism policies and critiques of industry and governmental principles (e.g. the UNWTO Global Code of Ethics for Tourism, which he contextualised as embedded in neoliberalism and characterised as anthropocentric in tone).

Other authors have investigated codes of conduct and codes of ethics relative to types of animal-based tourism experiences; for example, zoo management (Bahne, 2015), seal watching (Öqvist et al., 2018), turtle tourism, (Waayers et al., 2006) and boto (dolphin) tourism (Fung, 2021), to name a few. These codes are meant to guide visitor behaviour as well as industry behaviour towards a more ethical way of interacting with animals. And yet, even addressing ethical interactions with animals may not go far enough in terms of animal justice if the animals have been forcibly taken from their natural context or have little agency within their lives (Donovan & Adams, 2007).

It is here we must depart from the tourism literature to gain additional perspectives on what animal justice or species justice may look like. In their book *Zoopolis: A political theory of animal rights*, Donaldson and Kymlicka (2011) argued for animal rights based on a relational theory rather than a moral or ethical one. They began with the premise that:

> We are part of a shared society with innumerable animals, one which would continue to exist even if we eliminated cases of 'forced participation'...Ongoing interaction is inevitable and this reality must lie at the centre of a theory of animal rights, not be swept to the periphery.
>
> (Donaldson & Kymlicka, 2011, p. 8)

Animal "rights", in this case, is not absolute "non-intervention" but rather honouring our "positive duties" towards the animals we have domesticated as well as the uninvited creatures who interface with human societies. They posited that this means we must strive for a "just coexistence" with relationship-specific positive duties according to the animal at hand and where human activities and spaces are designed with animal interests in mind. In other words, they advocated for animal-centric thinking (Carr & Broom, 2018). Donaldson and Kymlicka criticised traditional animal rights perspectives as too narrow, unfeasible, politically unsustainable and ignoring the strong psychological pull for humans to connect with animals. They suggested grounds where the human-animal relationship "integrates universal negative rights *(i.e. non-exploitive relationships)* owed to all animals with differentiated positive rights depending on the nature of the human-animal relationship" (Donaldson & Kymlicka, 2011, p. 10). What is particularly unique about their approach, however, and what connects it to the area of socialising animals, is their positioning of animals within a political framework. They focus on the idea of citizenship, a concept that allows for broad individual rights to be combined with numerous situations and human-animal associations, impelling us to seriously consider the "broad range of issues about membership, mobility, sovereignty, and territory" and bear in mind that "many of the processes that generate the need for a group-differentiated theory of human citizenship also apply to animals" (Donaldson & Kymlicka, 2011, pp. 11 & 14).

Transitioning to the execution of just interactions between human societies and animals, O'Neill (1997) argued for obligation-based reasoning to protect the natural world, which she advised held more sway than realism, utilitarianism and rights-based reasoning. Obligations propel a type of behaviour-centred ethics focusing on actions rather than desired outcomes, which go further than arguing *who* has *what type* of rights. She contended:

> Although rhetoric of rights has become the most widely used way of talking about justice in the last fifty years, it is the discourse of obligations that addresses the practical question who ought to do what for whom? The anthropocentrism of rights discourse is, as it were, the wrong way up: it begins from the thought that humans are claimants

> rather than from the thought that they are agents. By doing so it can disable rather than foster practical thinking.
>
> (O'Neill, 1997, p. 132)

Shifting from a rights-based to an obligation-based discourse focuses our thinking on what human considerations should be towards animals rather than continue to argue over what type of rights animals "deserve", a discussion that inevitably would vary from animal to animal and culture to culture. O'Neill (1997, p. 132) emphasised the obligation-based discourse as the place to begin our conversation about animals, avoiding the tendency to "lead [the conversation] with the confused anthropocentrism of a rhetoric of rights" which inevitably leaves it "perennially vague just who is obliged to do what for whom". If we choose to dismantle our post-anthropocentric world, it could be the principles of Donaldson's and Kymlicka's *Political Theory of Animal Rights* and O'Neill's obligation-based reasoning that help us rebuild from our post-pandemic context.

The timing for this shift to obligation-based reasoning and a political theory of animal rights is upon us. Humanity's tragedies and travesties of the past are deep wounds; however, the world now faces several interlocking existential crises, including the rise of virulent nationalism, pervasive outcomes from climate change and critical vulnerabilities in public health. Within the intersection of these crises, we see many areas where animals are affected and that relate to tourism, including:

- Humans hunt, eat and trap wildlife (bushmeat) for consumption; however, unsustainable demand has given rise to wildlife markets which have been demonstrated to provide grounds for disease to spread among wildlife. Some of the unsustainable demand is driven by restaurants in urban areas as well as other supply chains for animal hides that are shaped to become shoes, hides, wallets (de Franco & Marcaccuzco, 2018; Keul, 2018).
- Zoonotic diseases are passed from non-human animals to humans. Human travellers then spread the disease further.
- As we face the sixth mass extinction of wildlife, the loss of any species threatens the balance of our ecosystems. However, wildlife trafficking of both endangered and threatened species for medicinal (e.g. bear bile, tiger blood, pangolin scales) and culinary (pangolin meat, shark fin soup) consumption is linked to tourism.
- Unsustainable, unhygienic and unethical farming of non-endangered animals to produce culinary ends (cattle raised for beef, pigs raised for pork, geese raised for foie gras, high demand fish and crustaceans caught or raised for "seafood") degrades the environment and risks the health of those who consume it.
- Farmworkers and slaughterhouse workers are forced to work in poor conditions that may ultimately spread diseases such as COVID-19.

- Varied types of visitor merchandise come from animal parts; some are endangered. Included in these are not only well-documented trophy kills but also smaller and less overt souvenirs such as live animal keychains.
- Animal dependency on tourists leads to lack of care and food during the COVID-19 crisis (Fischer, 2021a).
- Natural resource resilience and rewilding occurring in pockets of the world where human activity decreased during the pandemic (Crossley, 2020).

In order to achieve different results, we must resist continuing to use exploitive approaches and we must abandon systems that encourage unfettered capitalism. Socialism as a political reality has given rise to corruption in many nations, as has capitalism. But as a theory, socialism is a framework for organising and managing the resources of the world, such that production, distribution and exchange should be owned or regulated by the community as a whole. Cohen (2009, p. 80) asserted that a "socialist aspiration is to extend community and justice to the whole of our economic life". The principles of socialism are egalitarianism and a focus on community, but as Cohen (2009) asked, do these principles make socialism desirable and are they feasible? So, we must address what would the prospect of socialising animals in tourism look like by asking if it is desirable and if it is doable. For example, it is possible that a socialised approach to animal tourism would have its own type of localised exploitation of animals, since socialism is a framework typically meant to manage resources (e.g. animals) for people. Therefore, the socialist approach alone addresses only the human equity concerns of the moment. However, combining socialism with post-humanism may yield a justice framework that encompasses our obligations to animals in thinking about their citizenry considerations as well (as per Donaldson & Kymlicka, 2011). Finally, we will discuss the practical considerations of how to create, adapt and enact these ideal tourism projects using a model known to practitioners and academics – community-based tourism (CBT).

Post-humanism and socialism

Post-humanism has yet to be widely considered within tourism (Cohen, 2019); however, if adopted, it could challenge the negative "ism's" noted above. Post-humanism is a branch of contemporary philosophy that redefines the position of humans within the world, blurring the boundaries between humans and the natural world and surmounting what has been thought about our abilities and natural right as leaders (Danby et al., 2019). Additionally, post-humanism debunks our simplified categorisations of the world, in particular dualities and dichotomies ultimately constructed for human benefit. Relative to the topic of this chapter, post-humanism defines the human-animal divide as a "political construction, facilitating the human exploitation of animals" and "denies primacy to any component, including the human" (Cohen, 2019, p. 418). Lindgren and Öhman (2019, p. 1201) denote post-humanism as a "deconstructive movement of human superiority that aims to reveal alternative perspectives" of animal-human interactions.

Humanism, by definition, is anthropocentric; therefore, to transcend its rigid, hierarchical, "man"-centred framework is to be open to new ways of viewing the world and thereby new ways of approaching our encounters with and within it. Within tourism, the application of a post-humanist approach to business-as-usual activities would disrupt power structures, abolish the many forms of exceptionalism and redistribute natural resources, wealth and agency into a flat (versus hierarchical), more fluid and localised locus of control. The industry's consideration of animals would radically shift to an animal-centric (Carr & Broom, 2018), more-than-human (Dashper, 2020) and trans-culturally constructed (Bertella, 2018) consideration. As Bertella (2018) observed in her ecofeminist deconstruction of human views towards seals in Norway, the way we view and treat animals is inextricably tied to our self-perception and representation. However, the simultaneous, multiple and conflicting ways in which we view animals, and ourselves, lends particular difficulty to an adoption of a post-humanist view, even if we were to agree on it in principle.

One of the early applications of post-humanism to animal-based tourism was offered by Venegas and López-López (2021) and López-López and Venegas (2021), who adapted a post-humanist critique of two forms of cultural-heritage-laden animal performances in Mexico, the "zonkeys" of Tijuana (donkeys painted to look like zebras) and the practice of bullfighting, respectively. Additionally, Monterrubio and Pérez considered the horse-human relationship, concluding that horses serve our speciest "entertainment, economic and cultural purposes" and that "a complete embracement of post-humanism is needed to dissolve basic horse-human dichotomies" (2020, p. 1).

These treatments of animals as objects for show and performance certainly run contradictory to post-humanism; however, as Cohen (2019, p. 422) posited: "as posthumanism tends to encourage mutual involvement, rather than separation, between humans and animals, it might approve of soft interaction of tourists with animals insofar as these are not detrimental to the latter". Likewise, Lindgren and Öhman questioned the "subjectivity" of animals by humans, in that our sense of "the animal other" only serves to undergird our convinced notion that we are something different from animals:

> …why is subjectivity an important issue when it comes to human-animal relationships? First, although subjectivity has been a criterion for who is morally accounted for, 'otherness' and 'difference' have been described as its inferior counterpart (Braidotti, 2013). According to Plumwood (2002), a clear subject/object division could deny the agency of the one studied. Therefore, the separation of the human subject (the knower) and the nonhuman object (the known) reproduces the idea of

> nature as a non-ethical and non-agential sphere. Consequently, when nonhuman animals are deprived of their ethical and political agency, they can easily be instrumentalised, made profitable and commodified into different forms, such as nutrition (meat and dairy products) or as objects for infotainment (circuses, zoos and safaris).
>
> Therefore, our concern is the notion of 'animal otherness' when it is understood as a passive object to be exploited by the human subject. Consequently, a nomadic subjectivity is important in order to confront/disrupt situations where the 'animal other' is viewed as non-agential, or becomes an unquestionable object for human manipulation.
>
> (2019. p. 1204)

From an ontological standpoint, it is within post-humanism that we find a platform for species justice. Guia (2021), who also embraces nomadic subjectivities, delineates *types of responsibility, solidarity and advocacy* that he deems as the parameters of justice tourism. According to Guia, duty ethics – which promotes social liberalism and prescribes a social responsibility, a social solidarity and a cultural advocacy – offers us *sustainable tourism*. Ethics of care – which promotes social humanitarianism and prescribes a relational responsibility, an affective solidarity and a humanitarian advocacy – offers us *humanitarian tourism*. But it is affirmative ethics, which is grounded in post-humanism and advocates a political responsibility, a political solidarity and a political advocacy, where we reach *justice tourism*. Guia points out that it is the ethics of utilitarianism that leads to the neoliberal representations of tourism we have today. The framework of affirmative ethics he described helps to distinguish between the various labels of alternative tourism such as responsible tourism, solidarity tourism, advocacy tourism and sustainable tourism. Guia explained:

> Taking posthumanism seriously means actively resisting the co-optation of tourism by the market, that is, learning to contest neo-liberalism with others ... it paves the way for re-introducing political responsibility, solidarity and advocacy as positive world-making practices with which to subvert the current commodification and de-politicization of all forms of tourism.
>
> (2021, p. 517)

Reflecting back on the arc of tourism, Binkhorst and Den Dekker (2009) denoted the characteristics of "first and second generation experiences", describing first-generation experiences as staged entertainment and fun while the second-generation experiences were based on co-creation and more individualised to the tourists' background and values. What I am proposing is a third-generation experience that transcends our individual impulses and societal norms to a post-humanist and ethical lens – where we put our desires second in order to contribute to a common good. This is a problematic perspective for those living in Western, individualistic cultures as it expressly contradicts the rugged pioneer, free-market, and "if-you're-not-moving-ahead-you're-falling-behind" narratives etched into societal norms. Therefore, because of socialism's current lack of traction, a third-generation experience that puts others above or at least at equal footing with ourselves could gain traction as a moral road to take while traveling outside your home region.

We must conceive of tourism models that embody a post-humanist approach, grounded in affirmative ethics, while enacting a socialist aspiration, if greater justice is our goal. It seems pragmatic to start with a form of tourism that seems close to what we want to achieve, or that seems to embody the principles of a post-humanist socialism.

Pushing current models towards post-humanism

Current forms of alternative tourism include CBT, appropriate tourism, responsible tourism, ecotourism, geotourism and sustainable tourism. One could argue that these models, while positive and well intentioned, do not go far enough. Each of these manifestations of tourism is often embedded within an undergirding of neoliberal capitalistic principles and Western-based supply chains. And until 2020, it was practically unimaginable that any other model would be viable; the alternative forms listed before were "the best we could do" in light of the runaway train of capitalism. As Higgins-Desbiolles noted, however, regarding the transformative shift of possibilities in the shadow of the COVID-19 pandemic:

> The revelatory power of this crisis is enormous. Ideas that are rocked by this pandemic and responses to it include: that globalisation is an unstoppable force; that we cannot unlink from the logics of continuous economic growth; that consumerism is the key to expressing our identity; and that neoliberal capitalism is the best system for organising and allocating resources.
>
> (2020, p. 613)

These logics have held sway at the animal-tourism interface as well, with animals having long been characterised as a "resource" and commodified as a means of profit-making, both inside and outside of the tourism industry.

Animals figure into current-day crises in multiple ways. First, the COVID-19 pandemic is speculated to be a zoonotic disease, likely spread via the commonplace practice of wildlife and bushmeat markets. If humans were more careful in their use, consumption and trade of wildlife, a raging virus might have been mitigated. Second, animals provide innumerable ecosystem services, which are increasingly threatened as the climate crisis grows more severe and species die out. Viewing the climate crisis from a strictly utilitarian lens, allowing the extinction of species to escalate cuts off one of the most valuable "resources" we have to repair the damage that's already been done. Wildlife is not only a metric to be used as indicators of biodiversity, but they do the work of undoing our destruction, namely destruction created by turning forests into grazing lands, oceans into swirling trash heaps and coastlines into resort enclaves. Third, while animals do not directly play into the increasing exhibitions of nationalism and racism around the globe, the politics of oppression and exceptionalism do (Harper, 2012). Groups who oppress other human groups are also likely to oppress non-human groups (Gruen et al., 2014; Roberts, 1997; Wrenn, 2015). As Young (1992) points out, the five faces of oppression are exploitation, marginalisation, powerlessness, cultural imperialism and violence. While it is beyond the scope of this chapter to explicate these elements of oppression, we should note here that we find them at the overlap between climate change denial, extreme nationalism and global health and hunger concerns. Future scholarship and advocacy should address these characteristics of animal oppression within tourism.

Because of the pandemic and the unexpected potential for "resetting" some of our recently entrenched consumeristic norms, we must quickly re-evaluate and act, if indeed we have the will to redevelop tourism practices based on different models. Examining the principles of CBT is a good place to start. CBT is an accepted model of tourism that bestows decision-making power about development and management to involved members of a municipality or region and directs the costs as well as the benefits of tourism to be shared among the citizenry. The objective of many CBT projects is revenue generation as well as establishment of conservation incentives (Tubey et al., 2019). CBT gained traction in the tourism literature in the 1990s alongside ecotourism, blossomed in the 2000s and is still currently being developed and refined.

CBT projects are manifested in various forms, according to the local assets and priorities. The tourism literature yields assessments of specific projects as well as frameworks for success. CBT raises particular issues for the animal interface with tourism, specifically wildlife tourism or conservation-based tourism due to problems that arise in human-wildlife conflict. As Tubey, Kyalo and Mulwa noted: "for an average local individual, wildlife tourism does not provide an equally profitable use of their land compared to livestock or agricultural production. Moreover, other wildlife induced losses are almost never adequately addressed" (2019, p. 91). Additionally, human, crop or property losses due to wildlife destruction or disease transmission are well documented (Tubey et al., 2019; Walpole & Thouless, 2005). Walpole and Thouless (2005, p. 123) noted that while most wildlife tourism occurs in protected areas or on private lands, "the real battlegrounds between people and wildlife are outside parks and private reserves where poor rural people coexist with wildlife"; few projects factor in the wide variation between individuals in the cost of living with wildlife. Certainly there are cases of acclaimed CBT where the economic benefits and decision-making powers are not held in the community but distributed among local political leaders and/or appropriated by external actors. As Tubey et al. explained: "for community conservation to be fully realized, protection of biodiversity, land use planning, mitigation of community-wildlife conflict, empowerment of the local community as well as use of traditional knowledge is a prerequisite" (2019, p. 91). However, communities must have adequate capacity to create experiences with commercial viability, including features of access, security, quality, uniqueness, marketing and pricing strategies. In some circumstances, without outside influences, the community members do not have access to capital and/or understanding of commercial risk and market realities (Walpole & Thouless, 2005). Finally, the linkages between wildlife conservation goals and tourism benefits must be strong, visible and consistently met.

Mayaka et al. (2019) distinguished between two types of CBT. The first focuses on the outcome of tourism development in a community, while the second approach focuses on the development of the community through tourism; in other words, using tourism as a means to the end of increasing community good. The key dimensions used to monitor CBT efforts over time are involvement, power/control and outcomes; when using these three dimensions as a tool to view a CBT trajectory, one can see which type is manifesting within the community (Mayaka et al., 2019). The second approach, which in its ideal form would include active reflection and participation by residents as well as systems for ensuring resident equity within the process, would align with how Higgins-Desbiolles (2020, p. 617) described socialised tourism as activity that is "responsive and answerable to the society in which it occurs".

In this ideal CBT scenario, the power and control dimension would be characterised not only by an internal locus of decisions, but continuous criticality around activities that might reflect a localised context. The outcomes would enhance the residents' quality of life, as defined by the residents, to include perhaps increased self-efficacy, social justice, social cohesion and freedom (Mayaka et al., 2019). Simultaneously, an eye must be kept to control the type of outcomes and extent to which these outcomes benefit external audiences as well (Curcija et al., 2019; Zielinski et al., 2020).

In 2017, Giampiccoli and Mtapuri proposed a classification system based on equity, education, empowerment, endogenous (i.e. local efforts) and the environment as criteria for rating CBT projects. Later, in their literature review encompassing CBT and animals, Giampiccoli et al. (2020) developed a framework based on four elements of animal-based CBT: management, human perspective, type of relationship and role of animals. The management element they cite is a balance between animal welfare, capacity-building and community members' livelihoods, whereas the perspective element is the orientation of the supply-side and demand-side persons towards animals.

To include animals within the discussion of a revamped, ideal form of tourism adds layers of ethical complexity. People's knowledge about, opinions of, values regarding and ethical starting points related to animals vary widely (Fischer, 2021b). Conservationists tend to worry about species at the expense of individuals, while animal welfare advocates allow for caging, breeding, killing, testing on and eating animals as long as their care meets a minimal standard (Donaldson & Kymlicka, 2011). Additionally, within the discussion of animal sentience, agency and welfare, an animal continuum exists whereby some animals are more interesting, valuable or appealing than others, and this value varies according to the cultural context the animals find themselves in. But regardless of these inconsistencies, if animals are regarded as part of the community's resources within CBT, as other types of natural resources are, it would at least – in theory – position them in a way so as not to be as vulnerable to *external* forces and practices.

Therefore, a form of CBT with a post-humanist ethos must stay true to various communitarian principles and commitments, including:

- Consider animals from a suspended perspective that isolates speciest biases and puts aside assumptions about the animal other;
- Accept that the needs of community ecosystems are as important as individual needs;
- Guard against strictly instrumentalist relationships, including those with animals;
- Strive to continually suppress the individual urge to dominate (nature), control (resources) and be in charge (of the outcome);
- Ascribe to collective decision-making best practices that include animal-centric thinking;
- Adopt a political theory of animal rights towards the end of obligation-based policies; and
- Continually improve models of CBT through iterative practice and consistent evaluation and accountability.

Related to policy and accountability, Hall (2011) used data in national reporting to the Secretariat to the Convention on Biological Diversity to investigate the approaches used to align with the Convention. In comparing four types of governance, he determined that 46% of contributing nations adopted at least a partial communities' approach that emphasised capacity-building and increased participation by community members. Another 26%–37% were adopting some market approaches (e.g. financial incentives and educational programmes for operators) and 38%–53% adopted some network approaches (e.g. coordination of national programmes, sectoral cooperation and collaboration with trading partners and neighbouring nations). The takeaway insight here is, from a pragmatic sense, no singular governance doctrine is ideal.

To that end, slow capitalism, regenerative economies and other new and hybrid forms of governance models are gaining traction as realities of climate change, biodiversity loss and mass extinctions become widely understood

(Fullerton, 2015; Sheldon, 2019). However, because the pandemic has emphasised our need to reconstruct our local and global economies, it does not mean that "all options are on the table". Advocating for local control is not the same thing as supporting full cultural relativism, which is easily condemned when noting cultural practices such as bride burning, lynching and in the case of animals, the *pajaan*, a Thai practice used to break the spirit of an elephant to bend him or her to human will. Rather, an "overlapping consensus on a reasonable political conception of justice" (Rawls, 1987, p. 2) "despite a diversity of doctrines ... may be achieved and social unity sustained in long-run equilibrium, that is, over time from one generation to the next" (Rawls, 1987, p. 5).

Creating a socialised system of animal tourism

To see what a socialised system of animal tourism would look like, we must look at some of the best examples we have. However, when conducting a search for exemplary cases of animal-based CBT, it was difficult to find truly egalitarian and community-based projects to acclaim. Examples of animal-based tourism lacking animal welfare protocols or environmental protections are relatively easy to find (for example, see Atanga, 2019; Walsh & Zin, 2019). More often, CBT projects offer a "mixed bag" of benefits and shortcomings. For example, Stone and Stone (2011) reported drawbacks from the Khama Rhino Sanctuary Trust in Botswana, including community disappointment about loss of cattle grazing, lack of communication with the community, meagre employment numbers and the slow progress of the project which did not record a profit for 16 years. However, rhino numbers have risen and there is hope for future improvements such as appointment of a community liaison officer and training local people in tourism.

Mayaka et al. (2018) report on model projects of wildlife conservation and an elephant sanctuary in Kenya offering successful CBT features of recurring community participation, leadership dynamics, local trusting relationships and highly contextualised management styles according to local cultural traditions. The authors took a deep dive in each CBT case to outline the egalitarian and community-controlled elements of the projects. Barbieri et al. (2020) assessed seven CBT projects in the Peruvian Andes relative to participants' understanding of sustainability and any impacts generated from the projects. They found many aspects of the projects in line with principles and outcomes of sustainability, especially when the CBT initiatives are in line with traditional agriculture practices, including but not exclusively around animal husbandry (e.g. chickens, guinea pigs, sheep) and trekking with llamas. However, they also found management and networking inefficiencies requiring improvement in order to increase the overall sustainability of the CBT projects.

In each of these examples of CBT involving animals, some best practices that align with socialist principles are illustrated; however, in addition to other elements needing improvement, none of the projects were born from a post-humanist ethic and therefore it did not appear that the wildlife or domestic animals involved were considered as anything more than a resource. Several projects reported conservation "wins"; however, having pro-conservation objectives is still distinct from approaching animals with an obligations-based reasoning or with the concept of species justice in mind.

However, in 2020, Lemelin chronicled the case of the Stingless Bees of Mexico. Regarding its CBT and socialist principles, the project had two key features: first, a hyperlocal focus on Melipona bee species and caretaking practices; and second, an emphasis on the sustainability of the meliponaculture over tourism activities. Regarding a post-humanist approach, conservation is constructed as central to the tourist experience and is enmeshed in five senses to understand the bees using more than our "rational" mind. Additionally, meliponaculture is tied to Mayan religious ceremonies, and while bees are still viewed as a "resource", they are an honoured one. While this case summary is offered here only superficially, it provides a snapshot of what socialising CBT might look like in practise. Further and deeper examinations of other projects to refine the concept are critical.

The stingless bee example offers inspiration for future analysis of CBT projects in the hope of fostering continual learning loops on post-humanist, socialised animal experiences that could be enacted in tourism. While an abundance of case studies exists on CBT projects, I am not aware of any critiquing them through a post-humanist, socialised lens. Developing best practices – or inventories of what does and does not work within particular cultural contexts – could serve the field well as we move into a post-pandemic world. Identification of existing organisations working within the animal welfare and animal rights spaces who subscribe to socialist principles or post-humanist approaches (in practise, if not in name) would also lend a start. Deconstruction of the strategies of tourism organisations engaged in classic social movement principles of consciousness-raising, networking and self-efficacy (McGehee et al., 2014) could provide methods for developing projects embedded in principles of community and egalitarianism. While it is doubtful that certain parts of Western societies will ever unclench their power and wealth built on principles of capitalism, we can and should advocate for those that support self-governing community decision-making and democratic distribution of assets. The construction of this chapter as an academic exercise has demonstrated that a socialised system of animal tourism could exist by starting with the well-researched model of CBT. Using post-humanism as a foundation for how to view our fellow species and employing a political theory of animal rights and obligation-based reasoning as a means for developing practical ways co-existing and co-creating with animals, perhaps we can "make kin" of the varying views that humans hold about animals (Haraway, 2016). At least within the swath of the population that holds animals dear but that disagree on *how much welfare* we afford them and *to what degree* they have any rights, these two approaches to human-animal relationships provide a place to begin holding discussions. Additional empirical work and theoretical speculation are needed of course to advance our research, our activism and our knowledge on practical steps.

Time to turn: there is no other way

In 2018, Debbage laid out a series of "turns" within recent tourism and geographies literature: cultural, neoliberal, evolutionary, innovation and entrepreneurship and sectoral turns. Related more to critical studies of tourism, we have seen a hopeful turn (Ateljevic et al., 2013), a moral turn (Caton, 2012), a reflexive turn (Cohen, 2013), the animal turn (Andersson Cederholm et al., 2014; Danby et al., 2019) and the justice turn. The animal turn has been slow to develop within tourism, given that seminal works such as *Animal Geographies* by Wolch and Emel was produced in 1998.

This chapter has offered a way to move forward, and a difficult one at that. I have proposed a new perspective on CBT, applying socialist principles (of community and egalitarianism) within a post-humanist philosophy. To execute this at a pragmatic level, it must occur project by project and region by region, with tourism researchers working with industry to shine a light on the successful elements of tourism projects as well as acting as advocates for true CBT principles. We must do this so that justice tourism will blossom in pockets around the world and so that biodiversity systems, species as well as individual animals will eventually thrive again to the benefit of us all.

References

Adams, C. (1991). *The sexual politics of meat.* New York: Lantern Publishing.

Andersson Cederholm, E., Björck, A., Jennbert, K., & Lönngren, A. S. (2014). *Exploring the animal turn: Human-animal relations in science, society and culture.* Lund, Sweden: Pufendorf Institute for Advanced Scholars, Lund University.

Atanga, R. A. (2019). Stakeholder views on sustainable community-based ecotourism: A case of the Paga Crocodile Ponds in Ghana. *GeoJournal of Tourism and Geosites, 25*(2), 321–33.

Ateljevic, I., Morgan, N., & Pritchard, A. (Eds.). (2013). *The critical turn in tourism studies: Creating an academy of hope* (Vol. 22). London: Routledge.

Bahne, R. (2015). Ethics and code of conduct in zoo management [unpublished thesis]. University of Applied Sciences, HAAGA-HELIA.

Barbieri, C., Sotomayor, S., & Gil Arroyo, C. (2020). Sustainable tourism practices in indigenous communities: The case of the Peruvian Andes. *Tourism Planning & Development*, 17(2), 207–24.

Bertella, G. (2018). An eco-feminist perspective on the co-existence of different views of seals in leisure activities. *Annals of Leisure Research, 21*(3), 284–301.

Binkhorst, E., & Den Dekker, T. (2009). Agenda for co-creation tourism experience research. *Journal of Hospitality Marketing & Management, 18*(2–3), 311–27.

Braidotti, R. (2013). *The Posthuman*. Cambridge: Polity.

Büscher, B., & Fletcher, R. (2017). Destructive creation: Capital accumulation and the structural violence of tourism, *Journal of Sustainable Tourism, 25*(5), 651–67.

Carr, N., & Broom, D. M. (2018). *Tourism and animal welfare*. Wallingford: CABI.

Caton, K. (2012). Taking the moral turn in tourism studies. *Annals of Tourism Research, 39*(4), 1906–28.

Cohen, E. (2019). Posthumanism and tourism. *Tourism Review, 74*(3), 416–27.

Cohen, G. A. (2009). *Why not socialism?* Princeton, NJ: Princeton University Press.

Cohen, S. A. (2013). Reflections on reflexivity in leisure and tourism studies. *Leisure Studies, 32*(3), 333–7.

Crossley, É. (2020). Ecological grief generates desire for environmental healing in tourism after COVID-19. *Tourism Geographies, 22*(3), 536–46.

Curcija, M., Breakey, N., & Driml, S. (2019). Development of a conflict management model as a tool for improved project outcomes in community based tourism. *Tourism Management, 70*, 341–54.

Danby, P., Dashper, K., & Finkel, R. (2019). Multispecies leisure: Human-animal interactions in leisure landscapes. *Leisure Sciences, 38*(3), 291–302.

Dashper, K. (2020). More-than-human emotions: Multispecies emotional labour in the tourism industry. *Gender, Work & Organization, 27*(1), 24–40.

de Franco, M. R. B., & Marcaccuzco, J. A. M. (2018). Examining the correlation between tourism and the international trade of peccary: Ethical implications. In C. Kline (Ed.), *Tourism Experiences and Animal Consumption* (pp. 58–72). Oxfordshire: Routledge.

Debbage, K. (2018). Economic geographies of tourism: One 'turn' leads to another. *Tourism Geographies, 20*(2), 347–53.

Donaldson, S., & Kymlicka, W. (2011). *Zoopolis: A political theory of animal rights*. Oxford: Oxford University Press.

Donovan, J., & Adams, C. J. (Eds.). (2007). *The feminist care tradition in animal ethics: A reader*. New York City: Columbia University Press.

Duffy, R. (2013). The international political economy of tourism and the neoliberalisation of nature: Challenges posed by selling close interactions with animals. *Review of International Political Economy, 20*(3), 605–26.

Fennell, D., & Sheppard, V. (2019). Tourism operators on trial: Pushing the animal justice agenda forward in tourism in spite of theory. *Travel & Tourism Research Association Canada 2019 Conference*. Retrieved 23 November 2020, from https://scholarworks.umass.edu/ttracanada_2019_conference/10

Fennell, D. A. (2019). The future of ethics in tourism. In E. Fayos-Solà & C. Cooper (Eds.), *The Future of Tourism* (pp. 155–77). Cham: Springer.

Fennell, D. A. (2020). Tourism and wildlife photography codes of ethics: Developing a clearer picture. *Annals of Tourism Research, 85*, 103023.

Fennell, D. A., & Sheppard, V. (2020). Tourism, animals and the scales of justice. *Journal of Sustainable Tourism, 29*(2–3), 314–35.

Fischer, B. (2021a). Who should feed wild animals during a pandemic? *Journal of Applied Animal Ethics Research*, 1–15. https://doi.org/10.1163/25889567-BJA10012

Fischer, B. (2021b). *Animal ethics: A contemporary introduction*. London: Routledge.

Fullerton, J. (2015, April). *Regenerative capitalism: How universal principles and patterns will shape our new economy*. Stonington, CT: Capital Institute.

Retrieved 15 December 2020, from http://capitalinstitute.org/wp-content/uploads/2015/04/2015-Regenerative-Capitalism-4-20-15-final.pdf

Fung, C. (2021). Pre-legislation policy and practice: Boto conservation and tourism on the Rio Negro in the Central Brazilian Amazon. *Journal of Sustainable Tourism (volume and issue forthcoming).*

Garrod, B., & Fennell, D. A. (2004). An analysis of whalewatching codes of conduct. *Annals of Tourism Research, 31*(2), 334–52.

Giampiccoli, A., & Mtapuri, O. (2017). Beyond community-based tourism. Towards a new tourism sector classification system. *Gazeta de Antropologia, 33*(1). https://digibug.ugr.es/handle/10481/44467

Giampiccoli, A., Mtapuri, D. O., & Jugmohan, S. (2020). Community-based tourism and animals: Theorising the relationship. *Cogent Social Sciences, 6*(1). https://doi.org/10.1080/23311886.2020.1778965

Guia, J. (2021). Conceptualizing justice tourism and the promise of posthumanism. *Journal of Sustainable Tourism, 29*(2–3), 502–519.

Gruen, L. (2014). The faces of animal oppression. In S.N. Asumah & M. Nagel (Eds.), *Diversity, Social Justice, and Inclusive Excellence: Transdisciplinary and Global Perspectives* (pp. 281–296). Albany: SUNY Press.

Hall, C. M. (2013). A typology of governance and its implications for tourism policy analysis. *Journal of Sustainable Tourism 19*(4/5), 437–457.

Haraway, D. J. (2016). *Staying with the trouble: Making kin in the Chthulucene.* Durham, NC: Duke University Press.

Harper, A. B. (2012). Going beyond the normative white "post-racial" vegan epistemology. In P. Williams Forson & C. Counihan (Eds.), *Taking food public: Redefining foodways in a changing world* (pp. 155–174). New York: Routledge.

Higgins-Desbiolles, F. (2008). Justice tourism and alternative globalisation. *Journal of Sustainable Tourism, 16*(3), 345–64.

Higgins-Desbiolles, F. (2020). Socialising tourism for social and ecological justice after COVID-19. *Tourism Geographies, 22*(3), 610–23.

Holden, A. (2019). Environmental ethics for tourism-the state of the art. *Tourism Review, 74*(3), 694–703.

Jamal, T. (2019). Tourism ethics: A perspective article. *Tourism Review, 75*(1), 221–24.

Keul, A. (2018). Consuming the king of the swamp: Materiality and morality in South Louisiana alligator tourism. In C. Kline (Ed.), *Tourism experiences and animal consumption* (pp. 179–92). Oxfordshire: Routledge.

Lemelin, R. H. (2020). Entomotourism and the stingless bees of Mexico. *Journal of Ecotourism, 19*(2), 168–75.

Lindgren, N., & Öhman, J. (2019). A posthuman approach to human-animal relationships: Advocating critical pluralism. *Environmental Education Research, 25*(8), 1200–15.

López-López, Á., & Venegas, J. (2021). *Animal dark tourism in Mexico: Bulls performing their own slaughter.* In J. Rickly & C. Kline (Eds.), *Exploring non-human work in tourism: From beasts of burden to animal ambassadors* (pp. 69–82). Berlin: DeGruyter Publishing Oldenberg.

Malloy, D. C., & Fennell, D. A. (1998). Codes of ethics and tourism: An exploratory content analysis. *Tourism Management, 19*(5), 453–61.

Mayaka, M., Croy, W. G., & Cox, J. W. (2018). Participation as motif in community-based tourism: A practice perspective. *Journal of Sustainable Tourism, 26*(3), 416–32.

Mayaka, M., Croy, W. G., & Cox, J. W. (2019). A dimensional approach to community-based tourism: Recognising and differentiating form and context. *Annals of Tourism Research, 74*, 177–90.

McGehee, N. G., Kline, C., & Knollenberg, W. (2014). Social movements and tourism-related local action. *Annals of Tourism Research, 48*, 140–55.

Monterrubio, C., & Pérez, J. (2020). Horses in leisure events: A posthumanist exploration of commercial and cultural values. *Journal of Policy Research in Tourism, Leisure and Events*. https://doi.org/10.1080/19407963.2020.1749063

O'Neill, O. (1997). Environmental values, anthropocentrism and speciesism. *Environmental Values, 6*(2), 127–42.

Öqvist, E. L., Granquist, S. M., Burns, G. L., & Angerbjörn, A. (2018). Seal watching: An investigation of codes of conduct. *Tourism in Marine Environments, 13*(1), 1–15.

Plumwood, V. (2002). *Environmental Culture, the Ecological Crisis of Reason*. London: Routledge.

Rawls, J. (1987). The idea of an overlapping consensus. *Oxford Journal of Legal Studies, 7*(1), 1–25.

Roberts, L. (1997). One oppression or many? *Philosophy in the Contemporary World, 4*(1/2), 41–47.

Ryder, R. D. (2004). Speciesism revisited. *Think, 2*(6), 83–92.

Ryder, R. D. (2017). *Speciesism, painism and happiness: A morality for the twenty-first century* (Vol. 47). Exeter: Imprint Academic.

Scheyvens, R. (2002). *Tourism for development: Empowering communities*. Harlow: Prentice Hall.

Sheldon, P. (2019). *Regenerative economy in tourism*. Keynote presentation made at the 60th International Conference on Tourism in Dubrovnik, Croatia, November 2019.

Sheppard, V. A., & Fennell, D. A. (2019). Progress in tourism public sector policy: Toward an ethic for non-human animals. *Tourism Management, 73*, 134–42.

Stone, L. S., & Stone, T. M. (2011). Community-based tourism enterprises: Challenges and prospects for community participation; Khama Rhino Sanctuary Trust, Botswana. *Journal of Sustainable Tourism, 19*(1), 97–114.

The Superior Human? (2012, 30 March). Documentary movie; directed by Samuel McAnallen. Australia.

Tubey, W. C., Kyalo, D. N., & Mulwa, A. (2019). Socio-cultural conservation strategies and sustainability of community based tourism projects in Kenya: A case of Maasai Mara conservancies. *Journal of Sustainable Development, 12*(6), 90–102.

Venegas, J., & López-López, Á. (2021). Working donkeys in northwestern Mexico: Urban identity and tourism resources. In J. Rickly & C. Kline (Eds.), *Exploring non-human work in tourism: From beasts of burden to animal ambassadors* (pp. 53–68). Berlin: DeGruyter Publishing Oldenbourg.

Waayers, D., Newsome, D., & Lee, D. (2006). Research note observations of non-compliance behaviour by tourists to a voluntary code of conduct: A pilot study of turtle tourism in the Exmouth region, Western Australia. *Journal of Ecotourism, 5*(3), 211–22.

Walpole, M. J., & Thouless, C. R. (2005). Increasing the value of wildlife through non-consumptive use? Deconstructing the myths of ecotourism and community-based tourism in the tropics. *Conservation Biology Series-Cambridge, 9*, 122.

Walsh, J., & Zin, K. K. (2019). Achieving sustainable community-based tourism in rural Myanmar: The case of river Ayeyarwaddy dolphin tourism. *Zagreb International Review of Economics and Business, 22*(2), 95–109.

Winter, C. (2020). A review of animal ethics in tourism: Launching the annals of tourism research curated collection on animal ethics in tourism. *Annals of Tourism Research, 84*, 102989.

Wrenn, C. L. (2015). Animal oppression & human violence. *Between the Species, 18*(1), 9.

Young, I. M. (1992). Five faces of oppression. In T. Wartenberg (Ed.), *Rethinking Power* (pp. 174–195). Albany: State University of New York Press.

Zielinski, S., Jeong, Y., & Milanés, C. B. (2020). Factors that influence community-based tourism (CBT) in developing and developed countries. *Tourism Geographies.* https://doi.org/10.1080/14616688.2020.1786156

12 Buen Vivir

A guide for socialising the tourism commons in a post-COVID-19 era

Natasha Chassagne and Phoebe Everingham

Introduction

The COVID-19 crisis has seen an existential decline in the global tourism sector. Strict lockdowns in almost every jurisdiction coupled with stringent restrictions on global movement have led to the tourism sector becoming one of the hardest hit industries of the crisis. While COVID-19 threatens to be around for some time, many jurisdictions between and within countries are reopening or planning to reopen, albeit to a decimated sector. At the same time, people suffering from "lockdown fatigue" are eager to engage in what has been coined "revenge travel", "referring to the ways travellers intend to take their revenge on the limitations and restrictions of 2020 by splurging on bigger and more trips in 2020" (Burfitt, 2020). Yet governments, tourists and tourism businesses alike realise that the scale and form of tourism post-COVID will be indefinitely different. The focus is now on how tourism destinations can revive a decimated global sector; however, as we argue throughout this chapter, reverting to the status quo is not an option for the sustainability of the planet, nor will it sustain the tourism commons for current and future generations. Moving forward, tourism consumption based on economic growth models that exploit the natural environment must change.

Briassoulis (2002) calls the natural, built and sociocultural spaces for tourism "consumption" of the "tourism commons", which like all commons are open to the possibility of overexploitation if growth exceeds the limits of carrying capacity. Given the potential social and ecological repercussions from an accelerated tourism bounce-back, we argue that degrowth of the tourism industry is necessary for the global tourism commons to be enjoyed by current and future generations. Indeed, the pandemic has given rise to visions for the future of travel that involve a social and environmental "reset" (Everingham & Chassagne, 2020).

However, Hall et al. (2020, p. 579) argue that tourism becoming more sustainable after COVID-19 "is not a foregone conclusion". This may result in a battle between proponents of more neoliberal type growth in tourism that goes beyond the planetary carrying capacity and proponents of a tourism

DOI: 10.4324/9781003164616-12

that is more localised and slower, with fewer tourism-dependent economies, and featuring more equitable distribution of benefits. Acknowledging the possibilities for resetting tourism in sustainable ways, we assert that the global tourism industry has already experienced a shift in ecological impact due to the unintended degrowth of the economy. Like Higgins-Desbiolles (2020a), we argue that post COVID-19, a social and ecological reset is not only timely but also critical to change the parameters of the tourism trajectory. A shift in priorities towards tourism that enhances the social and environmental well-being of host communities is needed to establish meaningful human connections between hosts and tourists (Chassagne & Everingham, 2019). Tourism must be "socialised" in a way that recentres it on the public good towards a future tourism that prioritises host communities, planetary limits and future generations (Higgins-Desbiolles, 2020a).

Socialising tourism is described by Higgins-Desbiolles (2020b) as "harnessing it for the empowerment and wellbeing of local communities". This necessitates taking care of the tourism commons in relation to the natural environment. We propose that the Latin American concept of Buen Vivir can act as a guide, a framework to socialise tourism for the good of humans and the planet. Buen Vivir can be loosely translated as the "good life", which refers to a life "in plenitude", in harmony with each other and nature and is an endogenous, biocentric and holistic approach to well-being and sustainability, "as much ecological as it is social" (Chassagne, 2021, p. 2). Taking an endogenous approach to sustaining the tourism commons, based on the principles of Buen Vivir it provides a value system that underpins how the commons can be socialised. It is antithetical to the neoliberal premise of self-interest and exploitation. Buen Vivir has its origins in Indigenous cosmology; however, in its current conceptualisation, it involves interpretations from non-Indigenous communities, academia and policy. As such, it is a plural concept with no single definition as it relies on contextualisation; however, it is guided by a set of core principles to attain social and environmental needs. Buen Vivir denounces economic growth as an indicator of well-being and honours the connection between society and nature. In prioritising ecological principles and community well-being, it questions notions of economic development that are based on exploitation of people and the environment. Under Buen Vivir, community is not just based on Western conceptions of humans but expands the idea of citizenship to the non-human world, evoking a particular ideal of "the collective". While the "human community" is a key actor in decision-making, the needs of the environment are considered as inseparable to the needs of society. In the context of tourism, an endogenous concept like Buen Vivir can help prioritise community needs and respect planetary limits through reconceptualising ideas of collective social and environmental well-being. Tourism as an alternative economic activity guided by *Buen Vivir* can help do so by acting as a guide to "rethinking tourism".

This chapter begins with a discussion of the problems with tourism when underpinned by neoliberal capitalism. We argue for a reset of the industry away from business-as-usual to avert a "tragedy of the tourism commons". We then discuss how socialising tourism under the principles of Buen Vivir can foster a transformation of the tourism industry for a post-COVID era that is not only more socially and ecologically just for host communities and the planet, but also develops an industry that can provide tourists and hosts with mutually beneficial deeper and more meaningful experiences. We propose that Buen Vivir can act as a guide for the tourism industry, tourist behaviour and policy decision-making. It is our premise that this approach may help in efforts to "reset tourism" by growing the social and environmental well-being of the global community rather than seeking economic growth to satisfy the desires of visitors driven by a neoliberal approach to tourism.

Neoliberalism and a tragedy of the tourism commons

Neoliberalism is guided in large part by what Adam Smith called "the Invisible Hand" (Bishop, 1995). This describes an ideology that contends the best outcomes occur by reducing government intervention in the market. Continuous economic growth is desirable and driven by people acting in self-interest for self-gain in a so-called "free market" economy (Bishop, 1995). These ideals promote and uphold the self-interest of individuals who seek to gain benefit for themselves as opposed to the benefit of society as a whole. In this context, the neoliberal tourism market is an extractive activity, necessarily bound up in models of economic growth and development that benefits a minority of individuals rather than the majority population (Chassagne & Everingham, 2019). Tourism underpinned by neoliberal ideals focuses on tourism as a form of consumption, where economic growth – coupled with value for the tourism dollar – trumps negative impacts on the environment and host communities. This exponential growth of tourism has led to overtourism, a situation where there are too many tourists, resulting in a loss of quality of life for locals and ironically a situation where the quality of "authentic" experience for tourists also deteriorates (Goodwin, 2017). Degrowing tourism and effectively socialising tourism through a Buen Vivir approach promises to remediate these ills, which we will discuss later.

The neoliberal tourism market has already struggled to grapple with the social and environmental impacts from rapid growth in the sector, owing in large part to greater wealth accumulation and consequential rise in disposable incomes. This has made tourism, and particularly international tourism, more accessible to a larger number of middle-income households, who are often visiting regions with high socio-economic disparity. These tourism mobilities are framed as being equitable within the context of neoliberal development models because of the emphasis on the "trickle down" economic impacts. However as critical tourism scholars have pointed out, despite the claims that tourism can help "develop" and "boost" countries out of poverty, tourism has largely failed to "trickle down" to communities most in need (Chambers & Buzinde, 2015; Scheyvens, 2011).

Too often, host communities do not receive enough sustainable and tangible benefits that enhance their social and environmental well-being; to the contrary, there are often adverse social and ecological impacts and resources (the commons) become depleted or spoiled. Moreover, social tensions are created because of economic inequalities. Indeed, tourism development can merely work to exacerbate existing inequalities by allowing greater capital accumulation amongst the wealthy elite while further impoverishing local residents, for example, through land displacement, rises in the cost of living for locals and other forms of systematic marginalisation (Scheyvens, 2011). As Scheyvens (2011) points out, these inequalities can only be addressed through changing existing power relations and a focus on equity rather than "growth", and in the case of socialising tourism under the principles of Buen Vivir, alternative forms of tourism that value the well-being of the community as well as the well-being of their natural environment. Valuing environmental (and majority rather than minority) human well-being can help protect the tourism commons for the public good.

In his 1968 essay "Tragedy of the Commons", Garrett Hardin (1968) described "the commons" as a shared resource which is inevitably exploited due to the self-interest of individuals acting for independent gain. This "tragedy" deepens with greater urbanisation and land use changes (Hall et al., 2020, p. 580), but also when previously unattainable communities open up to tourism as part of neoliberal development agendas. Pandemic emergences related to a rise in zoonotic diseases, like COVID-19, are considered a consequence of human encroachment on the surrounding natural environment for "development" (Guégan, 2020), compounding the tragedy, which Hall et al. (2020, p. 4) argue, in terms of tourism "represents the intersection of broader processes of urbanisation, globalisation, environmental change, agribusiness and contemporary capitalism".

Ecological degradation of the commons directly affects a community's social and environmental well-being. Alternative forms of tourism, which are small in scale and focus on collective benefit, are therefore necessary in the aim of socialising tourism. Briassoulis says:

> The change from small-scale, family-based to large-scale, corporate-based tourism development brings about changes in the demand for and management of land, water, and other resources that are more serious in countries with underdeveloped or non-existent land use and resource planning and management institutions.
>
> (2002, p. 1079)

In the case of tourism, when everyone acts according to self-interest, the focus is on the benefits to the individual traveller and the capitalist development operator as opposed to the host community and natural environment. When tourism commons are degraded, the tragedy is not only of the natural environment and living conditions of host communities, but also these sites are no longer appealing for tourists and are sometimes even closed off to visitors. This also has detrimental effects for the local tourism operators. For example, Maya Bay on the island of Phi Phi Leh, Thailand, made famous from the movie "The Beach", had roughly 170 visitors per day in 2008. By 2017, that number had jumped to 3,500 – a case of overtourism that subsequently led to extreme environmental degradation and closure of the site until 2021 (BBC News, 2019). Overtourism then is a "consequence of unregulated capital accumulation and growth strategies" in relation to marketing of places as "tourism commodities" (Milano et al., 2019, p. 554). Despite the touted economic benefits of tourism expansion, host communities tend to find themselves in "social and ecological deficit" (Milano et al., 2019, p. 554).

Recent attempts to reconcile the social and ecological impacts of tourism with a desire to continue growth have resulted in responsible and sustainable tourism approaches. Sustainable tourism is defined as "tourism which meets the needs of present tourists and host regions while protecting and enhancing opportunity for the future" (Butler, 1999, p. 10). While there has been a considerable push within the tourism industry and the academy towards "sustainable" and "responsible" tourism, there is still an underlying belief that this can be obtained within current capitalist frameworks. This is exemplified in the United Nations World Tourism Organisation's (UNWTO, 2015–30) Sustainable Tourism Development Agenda, framed by the 17 United Nations Sustainable Development Goals (SDGs), which pushes for tourism as a driver of economic growth. Yet tourism can never be truly sustainable under market-driven neoliberalism; what is needed is "more critical and inclusive transitions toward sustainability" (Boluk et al., 2019). As Hall (2019) points out, it is a mistake to believe that more effort and greater efficiency alone will lead to sustainable tourism. Instead, what we need is a rethinking of human-environment relations (Hall, 2019).

Sustainable tourism development as part of a neoliberal agenda merely focuses on managing the sociocultural resources of host communities to preserve economic growth and to perpetuate the aforementioned satisfaction of tourists' desire for consumption rather than securing the self-sufficiency of host communities and intra-intergenerational equity in relation to costs and benefits (Briassoulis, 2002; Higgins-Desbiolles et al., 2019). This does not go far enough to increase social and environmental well-being of host communities, and within these models, the needs of the tourist therefore take precedence over the needs of local communities. The impacts are both direct and indirect.

In a post-COVID era, when we now understand how ecological disturbances can both impact and be impacted by future pandemics, it is imperative that we seek a paradigm change for tourism, which is what an approach such as Buen Vivir proposes. Considering that unsustainable human activity led to the pandemic in the first place (Guégan, 2020), a return to business-as-usual is not an option. Now is the time for change, paving a necessary way forward for sustaining the tourism commons for the benefit of natural environments and for the public good.

Degrowth: the imperative to sustain the tourism commons

As environmental and social issues are exacerbated by neoliberal models of tourism development that focus on growth, we therefore argue for a degrowth of the sector, concentrating on strengthening more socially and environmentally just alternative forms of tourism that bring sustainability and well-being of host communities to the core of tourism activities. In the context of COVID-19, the unintended degrowth of the global economy as an outcome of slowed economic activities globally gives us the opportunity to reset and transform both socially and ecologically. The increase in zoonotic diseases and the probability of future pandemics means that, as Hall et al. (2020, p. 585) state, "the demands for more sustainable forms of tourism will not fade away". Therefore, the way these sustainable forms of tourism will be approached will be vital. Degrowing economies and socialising tourism through Buen Vivir, therefore, helps to ensure environmental and social well-being for the public good and tourism commons.

At this juncture, it is vital to point out that sustainable degrowth is not the equivalent of an economic recession. Degrowth does not mean a backwards slide or degression; rather, it seeks to enhance sectors that are more socially and ecologically sound. While many countries are indeed experiencing recession as a consequence of COVID-19, the degrowing of certain sectors of the economy such as fossil fuels, extraction and other polluting industries provides an opportunity to concentrate economic efforts on those sectors that can be beneficial to both people and planet, if managed correctly. Hall (2009, pp. 55–6), citing the 2008 conference Declaration of the First International Conference on Economic Degrowth, has defined the characteristics of degrowth in the case of tourism, which include quality of life, basic needs satisfaction, societal change, increased self-sufficiency, self-reflection (non-materialism) and equity. These characteristics are synchronised with the aims of Buen Vivir and point towards a more just vision for tourism than the one offered by the neoliberal market.

Degrowth, however, requires effective policy, particularly in creating the spaces for local communities to drive decision-making when it comes to tourism activities. Within this momentum for a social and ecological reset, we can avoid a "tragedy of the tourism commons" (Briassoulis, 2002) by empowering local communities to decide on their own needs, manage local resources and make local decisions, including in matters of tourism as an alternative economic activity. Briassoulis (2002, p. 1080) described this as "the subsidiarity principle calling for actions at the lowest possible level and for external intervention only when certain functions cannot be performed locally".

Degrowing the tourism sector includes but is not limited to reducing the impacts of transport, such as the aviation industry. Regulatory agents and policymakers need to seek ways to sustainably reduce tourist numbers in communities, while letting those communities lead the direction of tourism activities to meet their social and environmental needs. This is where degrowing other sectors of the economy and focussing on promoting more sustainable and alternative forms of tourism can have an impact. This includes, for example, slow tourism, decommodified forms of volunteer tourism and community-based tourism, where local communities receive the benefits of the tourism dollars and locals and tourists both benefit from deeper, mutually beneficial intercultural exchanges (Chassagne & Everingham, 2019).

Within the frame of socialising tourism, it will need to be approached in a novel way rather than pursuing business-as-usual; otherwise transformation will not be possible. Hall et al. warned:

> COVID-19 may provide an impetus for individuals to transform their travel behaviours, however transformation of the tourism system is extremely difficult…resilience research in tourism highlights the need to consider biodiversity conservation and climate change imperatives in combination with destination models that seek to reduce leakage, enhance wellbeing, and better capture and distribute tourism value.
>
> (2020, pp. 584–5)

The push towards opening tourism up so that it can quickly return to "normal" (or a version thereof) requires regulatory and policy expectations with environmental caveats that businesses must meet certain sustainability or climate change requirements (Hall et al., 2020). Without these regulations, the possibility of socialising the tourism industry to enhance social and environmental well-being of host communities diminishes. Policy and decision-making must play a major role in ensuring that tourism is managed in a socially and environmentally just manner rather than leaving it to the "market knows best" model. Buen Vivir does this through a plural approach that contextualises the needs of local communities – human and non-human – through a participatory democracy that means all actors play a role, though it is community-led. This allows local communities directly affected by tourism to manage tourism commons rather than seeking land privatisation to maximise economic growth.

Exploitation of the tourism commons through privatisation of public land for tourism purposes not only causes social tensions within a community, but also has undesirable ecological impacts. There are numerous examples of private tourism development leading to the depletion of local natural resources through pollution, waste issues and natural habitat loss when tourism development is left to the devices of the market economy without government regulations. In a tourism context, this is particularly ironic as it is often the environmental resources on which tourism is dependent, as mentioned previously in the case of Maya beach in Thailand. Higgins-Desbiolles (2011) outlined some of these issues around governance in a case study of the development of an "ecolodge" on Kangaroo Island, Australia. When governance relies on the pursuit of tourism development embedded within growth models, "trade-offs" occur between environmental conservation and "demands for economic growth through tourism", often to the detriment of natural environments and host communities (Higgins-Desbiolles, 2011, p. 553).

The Lake Malbena case in Tasmania is another case in point, and emblematic of the neoliberal tourism market's disregard of community benefits and environmental impacts in the name of tourism development. The Lake Malbena project involves the development of luxury tourism, accessed by helicopter and involving the construction of a helipad on Hall's Island on Lake Malbena in Tasmania's Walls of Jerusalem National Park – part of the Tasmanian Wilderness World Heritage Area (Saker & Pendlebury, 2020). The controversial development is being protested by local communities, and at the time of writing, the development proposal has undergone a federal assessment of its environmental impacts (Gough, 2020).

The Tasmanian Wilderness Society has brought attention to the risks this development will have on Tasmanian wilderness, including the Environment Protection and Biodiversity Conservation (EPBC) Act that listed endangered flora and fauna such as the Tasmanian wedge-tailed eagle and the Alpine Sphagnum Bogs and Associated Fens (*The Wilderness Society (Tasmania) Inc v Minister for the Environment [2019] FCA 1842*, 2019). The EPBC Act is, in part, intended to implement Australia's international obligations under the World Heritage Convention to protect World Heritage sites (*Court finds Commonwealth decision on Lake Malbena 'heli-tourism' invalid, 2019)*. The development promises little benefit to locals, but rather may result in adverse impacts to local communities, walkers and other users of the public commons. Despite ongoing community opposition, a ruling by Tasmania's Supreme Court in July 2020 upheld a previous decision by the Resource Management and Planning Appeal Tribunal to grant planning permission for the proposal (McClymont, 2020).

The Lake Malbena case is a prime example of the impacts of policy and decision-making on tourism commons. In addition, where biodiversity and ecosystems are impacted by human activity and tourism development without due regard to environmental well-being, catastrophic flow-on effects can arise. The ecological damage is done when the capitalist value is sought to be maximised rather than the common good. As Hardin (1968, p. 1244) explained, "We want the maximum good per person; but what is good? To one person it is wilderness, to another it is ski lodges for thousands". Tourism underpinned by neoliberalism privileges acts of short-term self-interest at the expense of the benefits that tourism could have for the greater good, both ecologically and socially.

If the COVID pandemic has highlighted anything in the past year, it is the importance of the well-being of the collective and the impacts individual decisions can make on entire communities and even countries. A combination of the accumulating adverse impacts of neoliberal tourism on the ecological commons, together with the social, economic and physical limitations and flow-on effects to tourism from COVID-19, has demonstrated the need for transformative change in the tourism industry to better fit within the "social and ecological limits of the planet" (Higgins-Desbiolles, 2020a) Therefore, it is not only the nature of the tourism industry that is being forced to change, but also the nature of tourism consumption. A Buen Vivir approach can help scale back that consumption and empower local communities to manage their own tourism resources.

Local communities, local businesses, local needs and local environments need to be prioritised over the desires of exogenous actors like multinational organisations, large global tourism businesses and the privileged elite. A reimagined tourism that benefits host communities and natural environments can be incorporated at the policy level, but this also requires change driven from the bottom up that is community led. To that end, a Buen Vivir approach offers a framework recentred on social and environmental well-being of host communities.

Buen Vivir and post-COVID tourism transformation: a concrete guide for protecting the tourism commons

To socialise tourism via Buen Vivir, transformation of the tourism sector must seek to capacitate, benefit and empower local host communities over individual desires for the tourist experience. Socialising tourism through Buen Vivir allows for deeper connections between the tourist and the host community and a "greater sense of reciprocity with the natural environment" (Everingham & Chassagne, 2020, p. 557), to "degrow" neoliberal tourism and other industries that cause social and ecological harm. It can therefore be a tool that can be used to guide tourism transformation as an alternative economic practice, but in a way that respects the limits and the physical carrying capacity of the environment (Chassagne & Everingham, 2019).

As Fisher (2019, p. 456) states, Buen Vivir as an alternative to neoliberal development has resulted in tourism alternatives that are "more ecologically responsible alternatives to mainstream development". This is an approach that broadens the parameters of how tourism is framed and imagined: away from consumption models predicated on growth at the expense of the natural environment and towards a tourism where local knowledges and needs are prioritised. In turn, this provides more meaningful experiences for the tourists themselves, because they are learning from the cultures they are visiting and transforming their worldviews.

A Buen Vivir approach is holistic, considering environmental needs equal to social needs, which can help avert the "tragedy of the tourism commons". In her book *Governing the Commons*, Elinor Ostrom (1990) describes how communities can manage public space and resources without top-down governance. Her method of community-led governance draws similarities to the concept of Buen Vivir which demands that communities identify and manage their own needs within ecological limits. When we think about transformation in tourism then, this is not just about transforming thought and theory, but it requires action. What might this look like? Here we provide a guideline for how Buen Vivir can be put into practice – to socialise the tourism commons.

With the need for concrete guidance on how Buen Vivir is put in practice, Chassagne (2018) identified 14 conceptual principles for Buen Vivir from a critical analysis of the literature. In her book (Chassagne, 2021), these principles are triangulated with empirical data from Cotacachi, Ecuador, to expand to 17 core common principles for a framework for Buen Vivir practice. In a closer look at how these are put into practice, alternative forms of tourism were identified as an alternative economic activity that can be implemented within a Social and Solidarity Economy (SSE). The SSE includes, for example, prioritising local produce and local businesses, prioritising fair trade practices and featuring locally run co-operatives and associations. These can also help provide more meaningful experiences for tourists themselves, by immersing them in the places they are visiting and bringing them closer to local communities, cultures and environments.

The principles of Buen Vivir most relevant for tourism practice are culture, SSE, self-determination, participation, reciprocity, healthy environment, community and holistic rights. Socialising tourism practice for Buen Vivir would involve community-led tourism activities guided by the relevant principles of Buen Vivir. Reciprocity refers to "relationships of service and reciprocity towards each other and towards nature" (2012, p. 2) and holistic rights that to not only include fundamental human-centred rights, but also those accorded to nature (Chassagne, 2021, p. 62).

The types of tourism are thus important in maintaining community-driven processes and Buen Vivir requires alternatives to be "small-scale, local and benefiting local communities" (Everingham & Chassagne, 2020, p. 562), upholding the principle of an SSE. The focus on collective social *and* environmental well-being means that tourism activity has minimal impact on the tourism commons, applying the principle of reciprocity and a healthy environment. Here the principle of holistic rights comes into play by ensuring that alternative tourism activities uphold a respect for nature that does not view it as a commodity to be exploited for economic gain and tourism growth.

Community-based tourism, as an alternative economic activity to exogenously identified neoliberal development practices that are more socially and ecologically damaging, allows communities to develop and manage the tourism processes. This focuses on the principles of culture, self-determination, participation and community:

> Community-based tourism has the potential to fortify both the natural environment and the cultural heritage. In that respect tourism would not be mass-tourism on a large-scale for economic development, but more holistically approached and tourists thus would not be 'left to their own devices', practices, consumption and behaviours as tourists would be guided by local worldviews.
>
> (Chassagne & Everingham, 2019, p. 11)

As Higgins-Desbiolles (2020a, pp. 617–8) states: "The problem with tourism under neoliberal globalisation is that the power of society to manage, control and benefit from tourism businesses operating in their communities is undermined because the market is outside their control". However, tourism under Buen Vivir means that decision-making for the types of tourism and associated businesses is put in the hands of the host communities directly impacted by tourism. Communities take back control of the social and ecological benefits – and impacts – of tourism through local management of tourism processes. That is largely because social and environmental well-being for Buen Vivir is driven by the needs of the community and its environment in a participatory democracy rather than exogenously decided in a top-down approach to development.

A participatory democracy integrates the decisions, needs and opinions of the community in question to achieve their social, economic and environmental development. The type of tourism built off a participatory democracy thus "supports the shift back to local economies which are built for and by local communities, moving those communities away from the notion of 'developed places' for the purposes of the neoliberal tourist market" (Chassagne & Everingham, 2019, p. 11). Although an approach like Buen Vivir is community-led and focused on community needs, it also requires political will and cooperation to ensure that local processes are not co-opted by the dominant discourses for growth-fuelled development (Chassagne, 2019) in the desire to return to "normal" in a post-COVID economy.

These threats can be addressed by actors at all levels by changing the paradigm within which tourism currently operates but will require policy action to steer the industry away from the old "normal" and towards a transformative future. The aspect of plurality in Buen Vivir is therefore a pivotal one for two reasons. First, it relates to the plurality of knowledge, which involves prioritising local knowledges for both managing local environmental resources to avoid exploitation of the commons and also in identifying what communities need. Second, it relates to the plural implementation of tourism as an alternative economic activity, requiring cooperation from all actors, but particularly strategic support from policymakers as facilitators in capacitating communities (Bold, 2007; Chassagne & Everingham, 2019). A participatory democracy is thus a necessary precondition for creating the spaces and structures to enable communities to take charge (Chassagne, 2021) and helps cement the principles of participation and self-determination.

In a post-COVID era, the call to return to business-as-usual must be trumped by the undeniable logic that the neoliberal growth approach has failed. To paraphrase environmental youth activist Greta Thunberg, the so-called "normal" was always a problem. By socialising tourism and re-imagining how it can be otherwise through the guidance of Buen Vivir, we can start to visualise that "another world is possible" and indeed "another tourism is possible" (TAAF, 2020), resetting tourism for a socially and ecologically just future.

Conclusion: resetting tourism and growing social and ecological well-being to protect the tourism commons

The COVID-19 crisis has changed the face of tourism, providing an unprecedented opportunity for a social and ecological "reset" (Chassagne, 2020) and subsequently an "opportunity for new economic thinking" (Everingham & Chassagne, 2020, p. 559). In this reset, there is the momentum for pushing forward a greater emphasis on collective well-being and the public good rather than just individual well-being. In the case of tourism, this would place the focus on the benefits to communities and their ecological and social limits, in turn providing more meaningful experiences for tourists through mutual intercultural exchange with host communities.

Prioritising the needs of host communities addresses the growing inequalities in vulnerable host communities, guiding the "tourism reset" to "build back better", so to speak. By "better", we do not refer to the neoliberal conception of linear economic progress through "growth" models, but rather reinstating social and ecological justice for the public good, promoting holistic connections between humans and nature. This also means reframing tourism objectives in the search for "emancipatory solutions" (Higgins-Desbiolles, 2020a) in the fight against the commodification of people and places and of privatisation of public commons for tourism.

Taking advantage of the momentum for this "reset" to reframe tourism will require a transformative approach to avoid the adverse impacts of an industry pressed to return to pre-COVID "normal" levels of growth. Socialising tourism using the concept of Buen Vivir can thus act as a guide for policy, local decision-making and tourist behaviour for alternative forms of tourism. This requires a plural approach which involves governments, communities and tourists and a contextual lens to implement Buen Vivir in relation to the diverse needs of human and non-human communities.

As Higgins-Desbiolles (2020a, p. 620) argues: "socialising tourism requires both thought and action to ensure that we do not fail at this moment to secure real transformation from the disaster that the COVID-19 crisis has brought". Buen Vivir provides the opportunity to marry thought and action in tourism as a practical economic alternative towards greater collective social and environmental well-being.

Taking this approach to tourism in a post-COVID-19 era would be mutually beneficial to both host communities and tourists by enhancing tourists' experience through deeper human connections, a move away from "over-tourism" and a greater respect for the host's natural and social environment. The focus of the benefits needs to begin with increasing the social and environmental well-being of the host communities – empowering locals and building capacity and resilience in the "new normal". A recent study on the economic effects of the COVID-19 shutdown in the tourism-dependent Pacific by Regina Scheyvens and Apisalome Movono (2020, n.p.) demonstrates that host communities there are using this time to "rest", "recharge" and consider what is important in life:

> This break has given us a new breath of life. We have since analysed and pondered on what are the most important things in life apart from money. We have strengthened our relationships with friends and family, worked together, laughed and enjoyed each other's company.

The time has come to (re)consider the kinds of futures that are beneficial for the global commons more generally; tourism has a big part to play in providing alternative economies. However, extractive forms of tourism under neoliberal growth models will destroy the tourism commons for short-term self-interest. Buen Vivir can thus be utilised as a guide to help reset tourism for a more socially and ecologically just future post-COVID.

References

BBC News (2019, 10 May). *Thailand: Tropical bay from "The Beach" to close until 2021.* https://www.bbc.com/news/world-asia-48222627#:%7E:text=A%20Thai%20bay%20that%20was,had%20severely%20damaged%20the%20environment)%20

Bishop, J. D. (1995). Adam Smith's invisible hand argument. *Journal of Business Ethics, 14*(3), 165–80. https://doi.org/10.1007/bf00881431

Bold, V. (2007, October/November). Cotacachi actualiza su plan cantonal de turismo. *Periódico Intag*, 40 (Cotacachi, Ecuador).

Boluk, K. A., Cavaliere, C. T., & Higgins-Desbiolles, F (2019). A critical framework for interrogating the United Nations Sustainable Development Goals 2030 Agenda in tourism. *Journal of Sustainable Tourism, 27*(7), 847–64. https://doi.org/10.1080/09669582.2019.1619748

Briassoulis, H. (2002). Sustainable tourism and the question of the commons. *Annals of Tourism Research, 29*(4), 1065–85. https://doi.org/10.1016/s0160-7383(02)00021-x.

Burfitt, J. (2020). Why 2021 will be the year of 'revenge travel' (October 1, 2020). *Escape.* Retrieved from https://www.escape.com.au/news/why-2021-will-be-the-year-of-revenge-travel/news-story/4001546ceb090b2522300c4aca81bd3c

Butler, R. W. (1999). Sustainable tourism: A state-of-the-art review. *Tourism Geographies, 1*(1), 7–25. https://doi.org/10.1080/14616689908721291

Chambers, D., & Buzinde, C. (2015). Tourism and decolonisation: Locating research and self. *Annals of Tourism Research, 5*(1), 1–16. https://doi.org/10.1016/j.annals.2014.12.002

Chassagne, N. (2018). Sustaining the 'Good Life': Buen Vivir as an alternative to sustainable development. *Community Development Journal, 54*(3), 482–500. https://doi.org/10.1093/cdj/bsx062

Chassagne N. (2019) *Buen Vivir as an alternative to sustainable development: The case of Cotacachi, Ecuador.* Thesis, in Swinburne University of Technology (Melbourne, Australia).

Chassagne, N. (2020, 26 March). Here's what the coronavirus pandemic can teach us about tackling climate change. *The Conversation.* https://theconversation.com/heres-what-the-coronavirus-pandemic-can-teach-us-about-tackling-climate-change-134399.

Chassagne, N. (2021). *Buen Vivir as an alternative to sustainable development: Lessons from Ecuador (Routledge studies in sustainable development)* (1st ed.). London: Routledge.

Chassagne, N., & Everingham, P. (2019). Buen Vivir: Degrowing extractivism and growing wellbeing through tourism. *Journal of Sustainable Tourism, 27*(4), 1909–25. https://doi.org/10.1080/09669582.2019.1660668

Court finds commonwealth decision on Lake Malbena 'heli-tourism' invalid (2019). *Environmental defenders office.* Retrieved 20 October 2020, from https://www.edo.org.au/2019/12/05/lake-malbena-heli-tourism-decision-invalid/

Everingham, P., & Chassagne, N. (2020). Post COVID-19 ecological and social reset: Moving away from capitalist growth models towards tourism as Buen Vivir. *Tourism Geographies, 22*(3), 555–66. https://doi.org/10.1080/14616688.2020.1762119

Fisher, J. (2019). Nicaragua's Buen Vivir: A strategy for tourism development? *Journal of Sustainable Tourism, 27*(4), 452–71. DOI:10.1080/09669582.2018.1457035

Goodwin, H. (2017). The challenge of overtourism. *Responsible tourism partnership* (Working Paper 4).

Gough, A. (2020, 13 October). Lake Malbena helicopter tourism proposal to face Federal assessment. *Environmental Defenders Office.* https://www.edo.org.au/2020/09/17/lake-malbena-helicopter-tourism-proposal-to-face-federal-assessment/.

Guégan, J. (2020, 13 April). Virus: quand les activités humaines sèment la pandémie. *The Conversation.* https://theconversation.com/virus-quand-les-activites-humaines-sement-la-pandemie-135907

Hall, C. M. (2009). Degrowing tourism: Décroissance, sustainable consumption and steady-state tourism. *Anatolia, 20*(1), 46–61.

Hall, C. M. (2019). Constructing sustainable tourism development: The 2030 agenda and the managerial ecology of sustainable tourism. *Journal of Sustainable Tourism, 27*(7), 1044–60. DOI: 10.1080/09669582.2018.1560456.

Hall, C. M., Scott, D., & Gössling, S. (2020). Pandemics, transformations and tourism: Be careful what you wish for. *Tourism Geographies, 22*(3), 577–98. https://doi.org10.1080/14616688.2020.1759131

Hardin, G. (1968). The tragedy of the commons. *Science, 162*(3859), 1243–8.

Higgins-Desbiolles, F. (2011). Death by a thousand cuts: Governance and environmental trade-offs in ecotourism development at Kangaroo Island, South Australia. *Journal of Sustainable Tourism, 19*(4–5), 553–70. http://doi.org/10.1080/09669582.2011.560942

Higgins-Desbiolles, F. (2020a). Socialising tourism for social and ecological justice after Covid-19. *Tourism Geographies, 22*(3), 610–23. DOI: 10.1080/14616688.2020.1757748

Higgins-Desbiolles, F. (2020b). Why we need to think about "socialising tourism" after Covid, *Tourism Ticker.* Retrieved 4 September 2020, from https://www.tourismticker.com/2020/09/04/why-we-need-to-think-about-socialising-tourism-after-covid/#.X1GEFOjydRY.twitter.

Higgins-Desbiolles, F., Carnicelli, S., Krolikowski, C., Wijesinghe, G., & Boluk, K. (2019). Degrowing tourism: Rethinking tourism. *Journal of Sustainable Tourism, 27*(4), 1–19. DOI:10.1080/09669582.2019.1601732

McClymont, M. (2020, 28 July). Lake Malbena tourism decision faces fresh legal challenge. *Environmental Defenders Office.* Retrieved 4 September 2020, from https://www.edo.org.au/2020/07/28/lake-malbena-tourism-decision-faces-fresh-legal-challenge/

Milano, C., Novelli, M., & Cheer, J. M. (2019). Overtourism and degrowth: A social movements perspective. *Journal of Sustainable Tourism, 27*(12), 1857–75. http://doi.org/10.1080/09669582.2019.1650054

Ostrom, E. (1990). *Governing the commons: The evolution of institutions for collective action.* Cambridge: Cambridge University Press.

Saker, C., & Pendlebury, E. (2020). (20–001) The Wilderness Society (Tasmania) Inc v Minister for the Environment [2019] FCA 1842. *Environmental Law Reporter, 39*(20–001/20–007), 2–3.

Scheyvens, R. (2011). *Tourism and poverty.* New York: Routledge.

Scheyvens, R., & Movono, A. (2020, 1 November). Traditional skills help people on the tourism-deprived Pacific Islands survive the pandemic. *The Conversation.* https://theconversation.com/traditional-skills-help-people-on-the-tourism-deprived-pacific-islands-survive-the-pandemic-148987.

Tourism Alert and Action Forum (2020). Taking back our communities from tourism post COVID. Retrieved from https://www.facebook.com/groups/TourismAlertAndActionForum/.

The Wilderness Society (Tasmania) Inc v Minister for the Environment [2019] FCA 1842. *Environmental Law Reporter, 39*(20–001/20–007), 2.

13 Socialisation at scale

Post-capitalist tourism in a post-COVID-19 world

Robert Fletcher, Asunción Blanco-Romero, Macià Blázquez-Salom, Ernest Cañada, Ivan Murray Mas and Filka Sekulova

Introduction

The COVID-19 pandemic has illuminated cracks in the global tourism industry that have been growing in the shadows for some time. Among the many problematic issues historically associated with tourism development in many places, the following are commonly highlighted: unsustainable levels of resource consumption and pollution, including greenhouse gas emissions; social problems, including the rise of movements critical of touristification, drug use and sex work; and lack of economic diversification beyond the sector in populous destinations, coupled with workers' exploitation via precarious low-wage jobs (see Mowforth & Munt, 2016). While the global lockdown precipitated by the pandemic has notably reduced environmental and social impacts in numerous destinations, at least in the short term, it has greatly exacerbated economic disparities in places dependent on tourism revenue that has all but disappeared due to ongoing (if also constantly oscillating) travel restrictions both within and between societies worldwide.

Yet crisis, as always, can also be opportunity. Consequently, a variety of commentators have highlighted the potential to use both the problems and breathing space created by the pandemic as a valuable chance to "reset" or transform the tourism industry to redress the various problems the pandemic has highlighted. In this way, the industry might be steered towards a more sustainable course if and when the pandemic recedes (see especially, Lew et al., 2020). Among such proposals are calls to refocus the industry away from hegemony of the private sector, and the "perverse" incentives this creates to both hoard profit and externalise social and environmental costs in pursuit of this profit, in favour of forms of collective ownership, management and decision-making that would allow such costs to be "socialized" (see, e.g., Transforming Tourism Initiative, 2020). At the extreme, some proposals call for tourism to move away from its historical status as a quintessential capitalist industry altogether in pursuit of an anti- or "post-capitalist" agenda.

DOI: 10.4324/9781003164616-13

Thus far, however, calls for tourism socialisation post-COVID-19 have largely focused on community-level initiatives, echoing a longstanding emphasis on pursuing sustainable tourism more generally via "alternative" forms of "community-based tourism" (e.g. ecotourism and similar activities) (see Mowforth & Munt, 2016). While this is of course important, critics have long pointed out that the global tourism industry remains dominated by processes and operations that are decidedly translocal in scale. If this majority activity is ignored via a focus on community-level sustainability, such critics warn, the overarching industry can never become sustainable (Hall et al., 2020). Hence, as with the pursuit of sustainable tourism generally, in envisioning a post-COVID-19 future, there is a need to explore how tourism can be socialised at different levels and scales, including but also transcending community-level initiatives.

This is our aim in the present chapter. We begin by more fully outlining current calls for reforming tourism in the wake of the COVID-19 pandemic. We then explore how such calls relate to a critique of tourism growth as a whole that had been mounting in the years preceding the pandemic lockdown, leading to demands to refocus the industry around a concerted programme of "degrowth" (Kallis et al., 2020). Given that tourism as a capitalist industry demands continual growth to maintain stability, we point out, such degrowth would of necessity entail forms of post-capitalism. Building on Erik Olin Wright's (2019) discussion of different post-capitalist strategies cohering in an overarching programme termed "eroding capitalism", via a series of case studies grounded in our collective prior research we explore the potential to enact post-capitalist tourism at a number of different scales simultaneously. We conclude by highlighting the need for greater attention to how different initiatives aimed at socialisation can be combined in a concerted programme of "eroding tourism" in relation to its status as a quintessential capitalist industry.

The potential for post-capitalist tourism

In highlighting the devastating impact of COVID-19 on the tourism industry both globally and in specific destinations worldwide, commentators point to a wide range of problems both past and present that will need to be addressed in a post-pandemic "reset" (Gössling et al., 2020; Lew et al., 2020). Yet even prominent industry proponents like the United Nations World Tourism Organisation (UNWTO) and World Travel and Tourism Council (WTTC) assert the need to work harder to mainstream "sustainability" across the sector in a post-pandemic world. Such advocacy, however, still by and large emphasises a need to restimulate tourism growth as the foundation of this recovery (see, e.g., UNWTO, 2020).

This belies the fact that in the years preceding the pandemic, a mounting focus of tourism critique concerned this growth itself and its impact in the form of what many labelled a crisis of "overtourism" in numerous popular destinations (Milano et al., 2019). Consequently, critics questioned whether the sustainable tourism ostensibly pursued by the UNWTO and other influential industry leaders could ever be achieved in the context of the growth-based capitalist economy these leaders also advocated. The UNWTO (2018), among others, responded predictably by claiming that growth itself was not the problem; it was merely how such growth was managed.

Yet critics contended that this was not nearly enough, asserting the need to go beyond merely working to make tourism growth sustainable, to instead refocus the industry away from pursuit of growth altogether (Andriotis, 2018; Fletcher et al., 2020; Hall et al., 2020a). This position builds on an overarching critique of sustainable development more broadly as contradictory when committed to growth, arguing that current levels of economic activity are already far beyond "planetary boundaries" (Chakraborty, 2020; Rockström et al., 2009). Consequently, achieving genuine sustainability – let alone achieving this while also addressing rampant poverty and inequality – may require a concerted program of *degrowth*: an overall equitable reduction in economic activity in the Global North that diminishes throughput to sustainable levels while enhancing well-being (Kallis, 2018).

The degrowth proposal has been subject to a range of interpretations, elaborations, critiques and debates that are beyond the scope of this discussion (but see, e.g., Kallis et al., 2018, 2020). Suffice it to point out here that one important line of critique questions whether degrowth could ever occur within the context of a capitalist economy dependent on incessant growth to overcome internal contradictions (Foster, 2011; Liodakis, 2018). Consequently, degrowth advocates increasingly acknowledge that "[g]rowth is part and parcel of capitalism", and hence that "abandoning the pursuit of growth requires a transition beyond capitalism" (Kallis, 2018, p. 163).

Discussion of the potential to pursue degrowth in tourism therefore also foregrounds the question of whether this could be pursued within the capitalist economy or whether it requires pursuit of post-capitalism. Thus Büscher and Fletcher ask how "might tourism look if conceptualised from the point of view of a more general anti- or post-capitalist politics"? and answer by asserting that "tourism should move radically from a private and privatising activity to one founded in and contributing to the common" (2017, p. 664). Elsewhere, Fletcher expands on this to suggest that a proper post-capitalist tourism would pursue:

> (1) forms of production not based on private appropriation of surplus value; and (2) forms of exchange not aimed at capital accumulation; that (3) fully internalise the environmental and social costs of production in a manner that does not promote commodification and (4) are grounded in common property regimes.
>
> (2019, p. 532)

In terms of implementation, Higgins-Desbiolles et al. (2019) contend that touristic degrowth should privilege action at the community level first and foremost. Thus, they propose a "community-centred tourism framework that redefines and reorients tourism based on the rights and interests of local communities and local peoples" (Higgins-Desbiolles, 2020, p. 610). Reiterating this need for a community-level industry reorientation in response to the COVID-19 crisis specifically, Higgins-Desbiolles adds that this reorientation should entail tourism being more explicitly "socialised", explaining: "An agenda to socialise tourism would reorient it to the public good" (2020, pp. 610, 618).

This focus on socialisation of community-level and -led initiatives is indeed necessary and appropriate to support a post-capitalist degrowth touristic "reset". But is it enough? Will such community-level initiatives be sufficient to precipitate such a reset more broadly? A longstanding debate concerns whether sustainable tourism should similarly be pursued by refocusing activity towards community-level initiatives or whether one must also work to transform the larger-scale enterprises of which the global tourism industry is primarily composed (Butler, 1999; Hall et al., 2020; Mowforth & Munt, 2016). By the same token, should a plan to socialise tourism for post-capitalist degrowth also promote socialisation beyond the local level? And if so, how might this be achieved?

Eroding tourism?

These questions resonate with growing debate concerning the appropriate scale of focus within discussions of degrowth more broadly. Fraser (2013; Fraser & Jaeggi, 2018) asserts that in contrast to the conventional Polanyian "double movement" (Polanyi, 1944) in which social movements are seen to provoke a state response promoting social protection, degrowth should be understood as a "triple movement" pursuing an autonomous, emancipatory politics beyond state institutions. Consequently, as D'Alisa and Kallis (2020, p. 2) point out, thus far many "degrowth authors privilege bottom-up action by the grassroots" as their main scale for intervention. Yet as the authors also emphasise, many commentators then also "ask for top-down policy intervention from the state", given that their broader proposals often depend on state-level regulation, "without however offering a concrete view on the role of the state". Clearly, such analysts need to explain how such interventions could be achieved. Likewise, it is clear that Higgins-Desbiolles' (2020) own proposed policy measures for tourism socialisation include a number that would require active state-level intervention to be successful; yet she offers no guidance for how this could occur.

To direct attention to the potential for interventions to capture state processes and institutions in support of degrowth, D'Alisa and Kallis (2020) endorse a neo-Gramscian model of the state that views it not as a monolithic entity, but rather a constellation of different forces and interests that congeal into a particular hegemonic structure as certain points in time (see also, e.g., Jessop, 2016). Observing a similar ambivalence concerning potential for state action within anti-capitalist politics (which he equates with "*democratic* socialism") more generally, Wright argues that this is due to many activists' "belief that the character of the state in capitalist societies makes this impossible" (2019, p. 95). Contesting this position, Wright argues that while the state within capitalism is indeed usually co-opted in service of the status quo, this does not mean that it "cannot potentially be used to undermine the dominance of capitalism as well" (2019, p. 98). This is because "the apparatuses that make up the state are filled with internal contradictions" while "functional demands on the state are contradictory", opening potential for the state's co-optation in the interest of anti-capitalism as well (2019, p. 98).

Going further, Wright asserts that like the state, the capitalist system more broadly can be understood as a complex constellation of divergent processes rather than a coherent monolithic entity. From this perspective, he suggests: "the contrast between capitalism and socialism should not be regarded as a simple dichotomy"; rather, "we can talk about the degree to which an economic system is capitalist or socialist" (2019, p. 71). This perspective resonates with J.K. Gibson-Graham's (e.g. 1996, 2006) longstanding contention that "capitalo-centric" bias keeps us from acknowledging the diverse forms of economic activity that depart to greater or lesser degrees from capitalist logic existing within the interstices of the dominant system, a lens that others have productively used to highlight (the potential for) post-capitalist tourism practices in different contexts (see Cave & Dredge, 2020a; 2020b). Yet, more than Gibson-Graham, Wright maintains that capitalism continues to exercise an overall hegemony in most existing societies even if it harbours some post-capitalist spaces and potentials.

Based on this nuanced understanding of the nature of capitalism, Wright outlines a variety of specific strategies that can be understood as enacting anti- or post-capitalism in different ways and at different levels. *Smashing capitalism* entails efforts to overthrow the system as a whole, which Wright considers untenable, given that "evidence from the revolutionary tragedies of the twentieth century is that system-level rupture doesn't work as a strategy for social emancipation" (2019, pp. 41–42). *Dismantling capitalism*, by contrast, embodies the conviction that "a transition to democratic socialism could be accomplished through state-directed reforms that incrementally introduced elements of a socialist alternative from above" (2019, p. 42). *Taming capitalism*, Wright's third strategy, understands "capitalism as a source of systematic harms in society without attempting to replace it", but instead working "to build counteracting institutions capable of significantly neutralizing these harms" (2019, pp. 44, 45). Within this approach, *anti-capitalist* reforms are considered those "that introduce in one way or another egalitarian, democratic and solidaristic values and principles into the operation of capitalism" (Wright, 2019, p. 46). Wright's fourth strategy, *resisting capitalism*, comprises "struggles that oppose capitalism from outside of the state but do not themselves attempt to gain state power" (Wright, 2019, p. 49). *Escaping capitalism*, finally, embodies the contention that "[w]e may not be able to change the world at large, but we can remove ourselves as much as possible from its web of domination and create our own micro-alternative in which to live and flourish" (2019, p. 51). This is, of course, the very sort of local-level action that degrowth advocates along with current proposals for socialising tourism tend to privilege.

While dismissing *smashing capitalism* as counterproductive, Wright suggests that his other four strategies are in fact synergistic and capable of being combined into an overarching strategy he terms *eroding capitalism*. This would entail action on different levels simultaneously, bringing together "the bottom-up, civil society–centered initiatives of resisting and escaping capitalism with the top-down, state-centered strategy of taming and dismantling capitalism" into a powerful whole (Wright, 2019, p. 58). In this way, Wright explains: "The strategy of eroding capitalism combines initiatives within civil society to build emancipatory economic alternatives in the spaces where this is possible, with interventions from the state to expand those spaces in various ways" (2019, p. 95).

Combining Wright's different anti-capitalist strategies with the different scales of intervention they pursue provides a productive framework through which to explore possibilities for post-capitalism within tourism provision specifically. In the remainder of this chapter, we therefore adapt Wright's analytics to outline a series of case studies that can be understood to embody various combinations of these dimensions. Taken together, the cases can be understood as contributing to overarching strategy that we, following Wright, term "eroding tourism" in relation to the latter's conventional understanding as a capitalist industry *par excellence*.

Case studies

Barcelona: dismantling capitalism through municipal regulation

Spanish cities are being restructured and grown in accordance with the interests to commodify and financialise land development and the resulting built environment. Budgetary constraints and cliques of corruption have driven many urban and tourism planning decisions made by local authorities (Flyvbjerg et al., 2003). Tourist and real estate bubbles have been fed by the entrepreneurial management of the cities, aiming to compete for capital investment through place branding (Eisenschitz, 2016) and bypassing civil society in promoting urban megaprojects (Flyvbjerg et al., 2003). Within this context, tourist housing rental impacts on residential rentals and erodes the right to the city and its habitability (Martínez-Caldentey et al., 2020b).

In the face of such forces, the local governments of a few Spanish cities, such as Barcelona (Blanco-Romero et al., 2018), Madrid (Martínez-Caldentey et al., 2020a, 2020b) and Palma de Mallorca (Blázquez-Salom et al., 2019), have enacted measures restricting tourist rental in order to defend the right to housing as an extension to the right to the city (Madden & Marcuse, 2016). Such regulations respond to demands from social movements to decommodify urban life (Brenner et al., 2012), championed by governments that emerged from the 15M anti-austerity protests in 2011 and 2012 (Roth et al., 2019).

Barcelona (Catalonia-Spain) in particular has been, and continues to be, a laboratory of good and bad practices in tourism management. The Barcelona City Council charted a way out of the 2008 economic crisis centred on the development of a grand plan to promote the city as a prime tourist destination. This was so successful that conflicts generated by tourist overcrowding made tourism one of citizens' main concerns (Blanco-Romero et al., 2019). Starting in 2015, the City Council, governed by a coalition of left-wing parties (*Comuns*), started a new decision-making system focused on participation and citizen consensus. With the collaboration of civil society groups, the Council developed different post-capitalist measures for the management of the city as well as to contain commercialisation of key elements, such as homes transformed into tourist accommodations.

New intervention instruments that were approved included a fight against illegal tourist accommodation (mainly offered on online platforms like Airbnb). This began with the approval, in 2015, of a moratorium on the granting of licenses for the creation of tourist accommodations for a year. Despite its scope, this first moratorium had the initial support of the hotel sector and tourist apartments, until the arrival in 2017 of the Special Urban Plan for Tourist Accommodation (PEUAT, in its acronym in Catalan) regulating all tourist accommodations (including hotels, aparthotels, tourist apartments, pensions, hostels, houses for tourist use, student residences and youth hostels). The PEUAT was the first plan of its kind in Europe, designed based on extensive citizen participation (Blanco-Romero et al., 2018; Russo & Scarnato, 2017).

The main objective of the PEUAT is to improve the quality of life of the citizens of the city, with the aim of: (1) alleviating tourist pressure; (2) containing the increase in tourist accommodation; (3) preserving the quality of public space and diversify it with other activities; (4) promoting the diversity of urban fabrics; and (5) guaranteeing the right to housing, rest, privacy, well-being, sustainable mobility and a healthy environment. This pioneering initiative constitutes the regulatory framework for urban planning and management of tourist accommodation in the city through application of the urban planning law of Catalonia (Legislative Decree 1/2010). This decree regulates the creation of new tourist accommodation establishments and short-term housing rentals (STHRs). The PEUAT was designed as an urban planning instrument, dividing the city into four specifically regulated areas addressing and accounting for: the distribution of accommodation throughout the territory; the relationship between the number of tourist places offered and the resident population; the relationship and conditions of certain uses; the incidence of activities in the public space; and the presence of tourist attractions (Ajuntament de Barcelona, 2017). Like all planning and regulation exercises, despite the extensive public consultation, it has not met all of the demands of the myriad agents involved for various reasons. Yet it continues to be one of the few pioneering instruments in terms of both the creation process and the measures to be applied.

In parallel, trying to respond to the social conflict present in the city, the Strategic Tourism Plan 2020 was developed. This was one of the first intensive processes of reflection and networking on the tourist activity of the city and its effects. This effort focused on facing the key challenge of managing Barcelona as a tourist city compatible with the rest of the necessities of the multiple, complex and heterogeneous city that it is. Examples of this include the reorganisation of the port to move certain cruise terminals away from the centre and the prohibition of renting rooms for tourist use until the creation of a specific municipal regulation (Guerrero, 2020).

What this case illustrates is that regulation for post-capitalist tourism should focus not only on the "tourism issue," but on multiple dimensions of city management, to ensure protection against the various forms of social exclusion and "accumulation by dispossession" (Harvey, 2005) occurring throughout the territory in question. Barcelona thus endeavours to develop other public initiatives for the city's management not directly related to tourism, such as:

- The law approved by the Parliament of Catalonia for the limitation of rental prices;
- The purchase of buildings by the city council for their transformation into social rental housing, as part of the Plan for the Right to Housing 2016–25, thus combating gentrification and the expulsion of neighbours by investment funds and Real Estate Investment Trust (REIT);
- Creation of Points of Defence of Labour Rights, aimed at reversing dynamics such as job insecurity, economic monoculture or gentrification through the Barcelona Economic Development Plan 2016–21;
- Creation of energy advisory points where the City Council offers the necessary advice and support so that citizens at risk can exercise their energy rights before the basic supply companies;
- Promotion of a participatory process to decide the distribution of a part of the investments of the municipal budgets in the districts of the city within the project "Decidim Barcelona", initiated in 2016 and co-financed by the European Fund of Regional Development of the European Union.

Despite the institutional effort exerted to create this battery of initiatives, time is showing the limitations it faces. First, development of the proposed measures has been largely conditioned by the loss of control and institutional weight of the municipal government in the face of tourism industry promoter groups who continue to exert pressure to undermine these efforts. Additionally, lack of sufficient financing for the programme's full development leaves it without resources for implementation of important measures, and hence limits its effectiveness. This demonstrates the vital importance of the overarching governance process in deciding whether tourism activity can be designed according to other logics that do not focus on the reproduction of capital, but rather on objectives, principles and diverse development capacities oriented to the common good.

SESC Bertioga (Brazil): taming capitalism through social tourism

Since the 1930s and the 1940s, social tourism has been conceived as a means of providing access to free time to certain groups that could not enjoy it, mainly for economic reasons. With the Montreal Declaration of Social Tourism of 1996, and its addendum in 2006, this vision became more complex and also included in its objectives enhancing the well-being of tourism workers as well as the local communities and environment in which tourism operates (Schenkel, 2017). Additionally, recognition of more factors hindering access to tourism gave rise to programmes designed for the elderly, young people, people with different capacities, with serious diseases and in marginalised situations, among others (Minnaert et al., 2013).

Since its origin, social tourism has displayed contradictory and ambivalent dynamics from an emancipatory perspective (Minnaert et al., 2013). On the one hand, it supposes a kind of preventive social reformism, from which protective institutions of the working classes were promoted in order to distance them from trade union movements and class politicians, and therefore as a control and integration mechanism. But, at the same time, social tourism has also developed to institutionalise broad social demands, for instance, for the emancipation of female workers or for the needs of rest, well-being and personal development. On the other hand, its practices can be very different, even confrontational, ranging from initiatives fundamentally conceived as a way to subsidise the tourism industry, to programmes with a clear desire to train a critical citizenry that can fully develop their human capacities.

In this last sense, the experience of the Serviço Social do Comércio (SESC) São Paulo (SESC SP), founded in Brazil in 1946, by the entrepreneurs of the services, trade and tourism sectors, stands out (de Almeida, 2011; Schenkel & de Almeida, 2020). Its operation is regulated by a specific law and recognised in the Constitution since 1988, which has shielded it from various attempts to change its functions. It is financed through mandatory contributions from companies for this type of activity comprising 1.5% of all salaries paid. In its beginnings, the SESC SP had a strong welfare orientation and sought to address basic needs in matters related especially to workers' health. For this reason, they created hospitals and nurseries with the aim to improve people's hygiene and nutrition. But since the 1990s, SESC SP has been transformed into an institution oriented towards education in the non-formal sphere. Within this reorientation, social tourism was conceived as part of an educational action, with special attention to promoting artistic activities for the development of critical citizenship. This purpose is currently carried out through a network of 43 units in 21 cities.

Among these units, the SESC Bertioga stands out. It offers a beach resort in operation since 1948, aimed at workers in services, commerce and tourism (Cañada, 2020). At present, it can accommodate 1,000 people overnight, plus 350 who can enter daily to spend the day. Eighty-seven per cent of the people who stayed at the SESC Bertioga in 2018 were workers from these three sectors with incomes of between € 243 and € 1,215 per month (which makes this a destination for low- and medium-income workers rather than the truly destitute). The emancipatory potential of this initiative lies in the possibility of organising tourist activity that responds to the needs of a large majority of workers to access a coastal environment and a quality cultural and recreational offer that aims to enhance their capacities and social conscience. In addition, the working conditions of the initiative's own employees are of high quality, well above the norm for the hotel sector. Finally, instead of becoming an exclusionary enclosure, as all-inclusive resorts often do (Blázquez et al., 2011), SESC Bertioga can access the same educational, recreational and sports programmes as any other unit of SESC São Paulo, and hence offer these to its entire constituency.

FairHotel programme of the Unite Here Union (USA): taming capitalism through union organising

Job insecurity in the services industries is widely generalised (Antunes, 2020). In the tourism field, it has acquired a structural character, with a commitment to low labour costs on which a very prominent part of the industry's accumulation and reproduction strategy is based (Cañada, 2019). In this context, and despite often weak union organisation, collective action within the workplace constitutes one of the fundamental forms of control and imposition of social limits on these companies. Any post-capitalist transition can hardly be conceived in the short term, and therefore requires building solid alliances with trade union organisations and their workers that facilitate progress in a common direction.

Among the many union organisation experiences in the hotel sector, the North American union Unite Here has stood out for its mobilisation capacity in recent years. The union was founded in 2004 from the merger of two historic organisations, Here and Unite. Its focus is mainly on hotels, game rooms, food, laundry, transportation, airports, textiles and manufacturing and distribution sectors. It has nearly 300,000 members, mostly women and people of colour, with disproportionate representation by workers of immigrant origin. Thanks to union action, hotel workers, who have traditionally earned low wages, have managed to improve their salaries and access various other benefits. For example, the salary difference in hotels in Indianapolis, without a union presence, and Chicago, with unionised hotels, is clear (Cañada, 2017). Union organisation has also provided workers greater guarantees of rights in multiple areas, such as advice on immigration status and benefits in transportation, health, hours and work schedule, among others.

One of the most internationally recognised initiatives of this union has been the FairHotel programme. Through a webpage and an app designed for different mobile devices, anyone interested in staying in certain cities can access information concerning the quality of work conditions in different hotels. First, one can find an updated boycott list in which, ordered by states, one can locate hotels that are on strike, others in which some type of labour dispute is occurring and finally those that are at risk of entering into some type of labour dispute. Information is also provided, such that when a hotel is booked to stay or organise any type of event, one can secure a guarantee of protection against labour disputes, the occurrence of which allows for cancellation of reservation without penalty. In this way, the consumer can choose in an informed way in which hotels to stay when travelling or organising meetings, seminars or forums. The decision to boycott a hotel is taken by the unionised workers of the same hotel, constituting a tool for union pressure and action.

The programme also allows an interested person to be regularly informed about situations of labour conflict through an alert system. Likewise, various mechanisms allow those who desire to get involved in supporting the programme becoming something like permanent allies. In this way, private clients, organisations or companies that hire services in hotels can help to extend the number of hotels that request to be part of the programme or let these companies know that respect for labour rights is important in their policy via selection of hotels they want to work with.

The Programme is based on recognition of the importance of the opinion of clientele for the hotel sector. It works on the opportunity to promote those hotels offering decent working conditions and guarantees of free trade union organisation, and conversely, to penalise those companies in which labour rights are violated or that are in a conflict situation with labour. The extension of this type of alliance between diverse social sectors is fundamental to advance a broad post-capitalist project.

Conclusion

This chapter has explored how a post-capitalist tourism "reset" is or can be operationalised at scales beyond the community level. Three case studies have illustrated different strategies for contributing to such a project. The case of town hall policies in Barcelona, championed by governments that emerged from the 15M response, contributes to explorations of Wright's (2019) question of whether the state apparatus, usually co-opted in service of the capitalist system, can also be harnessed to undermine the latter's dominance. This SESC Bertioga initiative of social tourism in Brazil also illustrates a kind of preventive social reformism through protective institutions for the working classes, subsidised and regulated by the state. The FairHotel programme finally begins to demonstrate a more autonomous initiative directed by collective trade unions that pressures hotels to offer decent working conditions.

In their diverse forms and aims, this collection of cases supports Wright's contention that:

> [t]he optimal institutional configuration of a democratic-egalitarian economy is...likely to be a mix of diverse forms of participatory planning, public enterprises, cooperatives, democratically regulated private firms, markets, and other institutional forms, rather than to rely exclusively on any one of these.
>
> (Wright, 2019, p. 72)

The cases have all developed amidst the great contradictions of an environment dominated by capitalist relations, which cannot help but influence their form and outcomes. Taken together, they demonstrate the potential to combine diverse forms of action in different contexts and scales within an overarching strategy to *erode capitalism*, as Wright (2019) suggests. In this sense, different initiatives can be understood to contribute to a common "process of expanding and deepening the socialist elements of the economic system in such a way as to undermine the dominance of capitalism" (Wright, 2019, p. 71). As Wright describes the broader potential for such a strategy:

> Eventually, the cumulative effect of this interplay between changes from above and initiatives from below may reach a point where the socialist relations created within the economic ecosystem become sufficiently prominent in the lives of individuals and communities that capitalism can no longer be said to be dominant.
>
> (Wright, 2019, p. 62)

It is this potential to articulate touristic socialisation at scale that this chapter has sought to highlight and that we invite other researchers to explore further in the future.

References

Ajuntament de Barcelona (2017). Plan Especial Urbanístico de Alojamiento Turístico. Retrieved 2 March 2017, from: http://ajuntament.barcelona.cat/pla-allotjaments-turistics/es.

Andriotis, K. (2018). *Degrowth in tourism: Conceptual, theoretical and philosophical issues.* New York: CABI.

Antunes, R. (2020). As metamorfoses do mundo do trabalho e o proletariado de serviços/turismo. *Revista Turismo Estudos & Prácticas, 9* (Dossié Temático), 1–12.

Blanco-Romero, A., Blázquez-Salom, M., & Cànoves, G. (2018). Barcelona, housing rent bubble in a tourist city. Social responses and local policies. *Sustainability, 10*(6), 2043. https://doi.org/10.3390/su10062043

Blanco-Romero, A., Blàzquez-Salom, M., Morell, M., & Fletcher, R. (2019). Not tourism-phobia but urban-philia: Understanding stakeholders' perceptions of urban touristification. *Boletín De La Asociación De Geógrafos Españoles*, 83. https://doi.org/10.21138/bage.2834

Blázquez, M., Cañada, E., & Murray, I. (2011). Búnker Playa-Sol. Conflictos derivados de la construcción de enclaves de capital transnacional turístico español en el Caribe y Centroamérica. *Scripta Nova. Revista Electrónica de Geografía y Ciencias Sociales*, Universidad de Barcelona, *368*, 10/07/2011 (edición online).

Blázquez-Salom, M., Blanco-Romero, A., Vera-Rebollo, F., & Ivars-Baidal, J. (2019). Territorial tourism planning in Spain: From boosterism to tourism degrowth? *Journal of Sustainable Tourism, 27*(12), 1764–85. https://doi.org/10.1080/09669582.2019.1675073

Brenner, N., Marcuse, P., & Mayer, M. (Eds.). (2012). *Cities for people, not for profit.* London: Routledge.

Butler, R. W. (1999). Sustainable tourism: A state-of-the-art review. *Tourism Geographies, 1*, 7–25. https://doi.org/10.1080/14616689908721291

Büscher, B., & Fletcher, R. (2017). Destructive creation: Capital accumulation and the structural violence of tourism. *Journal of Sustainable Tourism, 25*(5), 651–67. https://doi.org/10.1080/09669582.2016.1159214

Cañada, E. (2017). Carrie Sallgren (FairHotel): "Creemos que hay mucho interés en poder viajar de una forma socialmente consciente". *Alba Sud.* Retrieved 1 March 2017, from http://www.albasud.org/blog/es/951/carrie-sallgren-fairhotel-creemos-que-hay-mucho-inters-en-poder-viajar-de-una-forma-socialmente-consciente

Cañada, E. (2019). Trabajo turístico y precariedad. En E. Cañada & I. Murray. *Turistificación global. Perspectivas críticas en turismo* (pp. 267–87). Barcelona: Icaria Editorial.

Cañada, E. (2020). *SESC Bertioga, donde el turismo social construye esperanza.* Barcelona: Alba Sud Editorial, colección Informes en Contraste, 11.

Cave, J., & Dredge, D. (Eds.). (2020a). *Reworking tourism: Diverse economies in a changing world.* London: Routledge.

Cave, J., & Dredge, D. (2020b). Regenerative tourism needs diverse economic practices. *Tourism Geographies, 22*(3), 503–13. https://doi.org/10.1080/14616688.2020.1768434

Chakraborty, A. (2020). Can tourism contribute to environmentally sustainable development? Arguments from an ecological limits perspective. *Environment Development and Sustainability.* https://doi.org/10.1007/s10668-020-00987-5

D'Alisa, G., & Kallis, G. (2020). Degrowth and the state. *Ecological Economics, 169*, 106486.

de Almeida, M. V. (2011). The development of social tourism in Brazil. *Current Issues in Tourism, 14*(5), 483–9. https://doi.org/10.1080/13683500.2011.568057

Eisenschitz, A. (2016). Tourism, class and crisis. *Human Geography, 9*(3), 110–24.

Fletcher, R. (2019). Ecotourism after nature: Anthropocene tourism as a new capitalist 'fix.'" *Journal of Sustainable Tourism, 27*(4), 522–35. https://doi.org/10.1080/09669582.2018.1471084

Fletcher, R., Murray Mas, I., Blanco-Romero, A., & Blázquez-Salom, M., Eds. (2020). *Tourism and Degrowth: Towards a Truly Sustainable Tourism.* London: Routledge.

Flyvbjerg, B., Bruzelius, N., & Rothengatter, W. (2003). *Megaprojects and risk: An anatomy of ambition.* Cambridge: Cambridge University Press.

Foster, J. B. (2011). Capitalism and degrowth: An impossibility theorem. *Monthly Review, 62*(8), 26–33.

Fraser, N. (2013). A triple movement? Parsing the politics of crisis after Polanyi. *New Left Review, 81,* 119–32.

Fraser, N., & Jaeggi, R. (2018). *Capitalism. A Conversation in critical theory.* Cambridge: Polity.

Gibson-Graham, J. K. (1996). *The end of capitalism (as we knew it): A feminist critique of political economy.* Minneapolis: University of Minnesota Press.

Gibson-Graham, J. K. (2006). *Postcapitalist politics.* Minneapolis: University of Minnesota Press.

Gössling, S., Scott, D., & Hall, C. M. (2020). Pandemics, tourism and global change: A rapid assessment of COVID-19. *Journal of Sustainable Tourism, 29*(1), 1–20. https://doi.org/10.1080/09669582.2020.1758708

Guerrero, D. (2020). Barcelona prohíbe el alquiler de habitaciones a turistas. *La Vanguardia.* Retrieved 26 August 2020, from https://www.lavanguardia.com/local/barcelona/20200826/483060699561/barcelona-alquiler-habitaciones-turistas-prohibido.html

Hall, C. M., Scott, D., & Gössling, S. (2020) Pandemics, transformations and tourism: Be careful what you wish for. *Tourism Geographies, 22*(3), 577–98. https://doi.org/10.1080/14616688.2020.1759131

Harvey, D. (2005). *A brief history of neoliberalism.* Oxford: Oxford University Press.

Higgins-Desbiolles, F. (2020). Socialising tourism for social and ecological justice after COVID-19. *Tourism Geographies, 22*(3), 610–23. https://doi.org/10.1080/14616688.2020.1757748

Higgins-Desbiolles, F., Carnicelli, S., Krolikowski, C., Wijesinghe, G., & Boluk, K. (2019). Degrowing tourism: Rethinking tourism. *Journal of Sustainable Tourism, 27*(12), 1926–44. https://doi.org/10.1080/09669582.2019.1601732

Jessop, B. (2016). *The State: Past, Present, Future.* London: Polity Press.

Kallis, G. (2018). *Degrowth (The economy: Key ideas).* New York: Agenda Publishing.

Kallis, G., Kostakis, V., Lange, S., Muraca, S., Paulson, S., & Schmelzer, M. (2018). Research on degrowth. *Annual Review of Environment and Resources, 43,* 291–316.

Kallis, G., Paulson, S., D'Alisa, G., & Demaria, F. (2020). *The case for degrowth.* Cambridge: Polity Press.

Lew, A., Cheer, J., Brouder, P., Teoh, S., Clausen, H. B., Hall, C. M., Haywood, M., Higgins-Desbiolles, F., Lapointe, D., Mostafanezhad, M., Mei Pung, J., & Salazar, N. (Eds.). (2020). Special issue on "visions of travel and tourism after the global COVID-19 transformation of 2020". *Tourism Geographies, 22*(3), 455–66. https://doi.org/10.1080/14616688.2020.1770326

Liodakis, G. (2018). Capital, economic growth, and socio-ecological crisis: A critique of de-Growth. *International Critical Thought, 8*(1), 46–65.

Madden, D., & Marcuse, P. (2016). *In defense of housing. The politics of crisis.* London: Verso.

Martínez-Caldentey, M. A., Murray, I., & Blázquez-Salom, M. (2020a). En la ciudad de Madrid todos los caminos conducen a Airbnb. *Investigaciones Turísticas, 19,* 1–27. https://doi.org/10.14198/INTURI2020.19.01

Martínez Caldentey, M. A., Murray, I., & Blázquez-Salom, M. (2020b). Habitabilidad y Airbnb: el alquiler de la vivienda en el distrito centro de Madrid. *Cuadernos de Turismo, 46*. https://doi.org/10.6018/turismo.451881

Milano, C., Cheer, J. M., & Novelli, M. (Eds.). (2019). *Overtourism: Excesses, discontents and measures in travel and tourism*. New York: CABI.

Minnaert, L., Maitland, R., & Miller, G. (eds.) (2013). *Social tourism. Perspectives and potential*. Abingdon: Routledge.

Mowforth, M., & Munt, I. (2016). *Tourism and sustainability: Development, globalisation and new tourism in the Third World* (4th ed.). London: Routledge.

Polanyi, K. (1944). *The great transformation: The political and economic origins of our time*. Boston, MA: Beacon.

Rockström, J., Steffen, W., Noone, K., Persson, Å., Chapin III, F. S., Lambin, E., … & Nykvist, B. (2009). Planetary boundaries: Exploring the safe operating space for humanity. *Ecology and Society, 14*(2), 32.

Roth, L., Monterde, A., & Calleja-López, A. (Eds.). (2019). *Ciudades democráticas. La revuelta municipalista en el ciclo post-15M*. Barcelona: Icaria.

Russo, A. P., & Scarnato, A. (2017). "Barcelona in common": A new urban regime for the 21st-century tourist city? *Journal of Urban Affairs, 40*, 455–74. https://doi.org/10.1080/07352166.2017.1373023

Schenkel, E. (2017). *Política turística y turismo social: Una perspectiva latinoamericana*. Ediciones CICCUS / CLACSO.

Schenkel, E., & de Almeida, M. V. (2020). Social tourism in Latin America: Regional initiatives. In A. Diekmann & S. McCabe (Eds.), *Handbook of social tourism* (pp. 33–42). Cheltenham: Edward Elgar Publishing Limited.

Transforming Tourism Initiative (2020, 27 September). Open letter: Covid-19 – now is the time to transform tourism (to UNWTO on world tourism day). Retrieved 15 October 2020, from http://www.transforming-tourism.org/fileadmin/baukaesten/sdg/downloads/Working_Programme/Open_Letter/Transforming_Tourism_Letter_to_UNWTO_01.pdf

United Nations World Tourism Organization (UNWTO) (2018). *'Overtourism'? Understanding and managing urban tourism growth beyond perceptions*. Madrid: UNWTO.

United Nations World Tourism Organization (UNWTO) (2020, 15–17 September). The Tbilisi Declaration: Actions for a Sustainable Recovery of Tourism. Retrieved 15 October 2020, from https://www.unwto.org/actions-for-a-sustainable-recovery-of-tourism.

Wright, E. O. (2019). *How to be an anticapitalist in the twenty-first century*. London: Verso.

Conclusion

Socialising tourism as an avenue for critical thought and justice: ways forward

Adam Doering, Bobbie Chew Bigby and Freya Higgins-Desbiolles

Introduction

This concluding chapter serves to summarise the varied contributions, key themes and future directions of *Socialising Tourism: Rethinking Tourism for Social and Ecological Justice*. This edited collection has sought to advance the socialising tourism concept by furthering ideas on how tourism may be made accountable to social and ecological limits. By bringing into conversation a diverse range of authors, academics, artists and activists already engaged in critical and often provocative analyses, the introduction and 13 chapters presented in this book offer an exploration into the possibilities of and impediments to the socialising tourism concept. Introduced by Higgins-Desbiolles in early 2020, the term "socialising tourism" was employed in this edited volume as a conceptual net cast out to gather and consolidate conversations concerning the potential for this historic transformative moment to reorient tourism towards more just and sustainable futures. The collection has taken the reader through a diverse range of countries and contexts. From Tar Creek to Tohoku, the book sought to improve our understanding of the possibilities, dynamics and limitations of the socialising tourism. The contaminated contexts and often violent cases selected for this book clearly showed that socialising tourism may not sit very comfortably within mainstream tourism discourse, practice and scholarship, but it is all the more important because of this. On a theoretical level, the book has been inspired by earlier critical work dedicated to approaching tourism as a social force, one significantly shaped by the incessant demands and growth imperatives of a capitalist political economy (Bianchi, 2009; Fletcher, 2011; Higgins-Desbiolles, 2006). Working across multiple scales, frameworks and logics, this collection opened up new lines of questioning and engagements to support the transformation of tourism as communities, governments and tourism businesses negotiate this rapidly evolving COVID-19 era.

In this concluding chapter, we reflect on what we have learned so far about the socialising tourism concept with respect to the three themes that structured the book: tourism social relations, tourism ideologies and suggestions

DOI: 10.4324/9781003164616-15

on how to build better collective tourism futures. This discussion highlights key emergent themes and points of tension, paying particular attention to the emphasis placed on the values of reciprocity, care and connection; the ongoing necessity of critique and collaboration; and the multiple scales and logics required to build alternative tourism futures. This is followed by a final discussion concerning future research and action agendas that aim to collectively build our capacity to socialise tourism for achieving social and ecological justice.

What we've learned

Rethinking tourism's social relations: reciprocity, care, connection

Section I of the edited volume explored the theme of rethinking social relations in tourism from a diverse range of contexts. In their own unique ways, each chapter examined how tourism's social relations between hosts and guests as well as people and their environments are significantly shaped by the political economy, class, race, gender and colonial histories. The discussions began by identifying the problematic social relations that define contemporary tourism, characterised by disconnection (Peters & Lambert, Chapter 1; Bigby & Jim, Chapter 2), carelessness (Carnicelli & Boluk, Chapter 3) and exclusion (Melubo & Doering, Chapter 4). These terms continue to characterise the experience of tourism for many communities who are confronted by international tourism and neoliberal social policies that invade and shape their living landscapes, waters and skies. Writing within and against these contexts, the chapters explored how communities carve out spaces of sovereignty, agency and autonomy, finding new ways to activate tourism for social and ecological justice. These contributions offer critical insights into how socialising tourism can help direct the industry and its practices towards new social relations that favour principles of reciprocity, care and connection.

The opening chapter by Peters and Lambert is exemplary of this approach, shedding light on the potential for Indigenous welcoming ceremonies to act as transformative tourism practice by bringing to the forefront Indigenous values of reciprocity, respect and connection that underpin the social relations of the Māori and the Aboriginal communities. Welcome ceremonies were shown to play a critical role in socialising tourism to better connect with the local peoples, lands and histories of the places they are visiting. Kajihiro's contribution in Chapter 8 makes a similar call to revalue or rethink tourism to favour Kānaka 'Ōiwi (Native Hawaiian) values of fostering relationships and mutual responsibilities over commercialised tourism engagements. The theme of connection also played a central role in Bigby and Jim's contribution in Chapter 2 where the authors further stressed the importance of connection for socialising tourism. Detailing the experience of toxic tourism in Tar Creek, the authors showed how such tours foster processes of connection at multiple levels, between visitors and local Indigenous communities as well as with the living waterways and landscapes where such tours take place. The transformative and educative potential of tourism was further emphasised by Carnicelli and Boluk's chapter (Chapter 3). Here, the authors argued that when incorporated into a critical pedagogy curriculum, a concept of care can be a vital pathway for socialising tourism towards social and environmental justice. It was suggested that building "caring capacity" through a critical pedagogy curriculum can play a significant and disruptive role for redirecting the future of the industry. Caring requires inclusion, and in Chapter 4, Melubo and Doering explored the idea that greater local participation in tourism *as* tourists—and not only hosts or stakeholders—could help reimagine and reconstruct Tanzanian tourism in a more culturally embedded and inclusive way. Collectively, these chapters showed us how approaching tourism based on values of reciprocity, care and connection has the potential to transform our relationships with one another, our shared histories and with the living lands and waters where tourism takes place.

Rethinking tourism ideology: on the necessity of critique and collaboration

Section II raised important questions about the concept of "socialising tourism" itself, which is appropriate given that the focus of this section was on a critique of tourism ideology. The section began with Bianchi's (Chapter 5) critique of the political economy and class relations in tourism. Bianchi offered a much-needed reflection on the growth-led, corporate-managed and resource-intensive ideologies, which he argued cannot be adequately addressed without attending to the contested class relations of power that fuel tourism's capital accumulation and inequality. Whereas Melubo and Doering (Chapter 4) suggested that a pandemic-driven market restructuring to favour domestic tourism markets in Tanzania could offer a much-needed shift towards a culturally inclusive form of tourism in the region, Bianchi made it clear that domestic markets are not easy substitutes for the resource- and finance-intensive international tourism markets. Put simply, Bianchi reminds us that existing tourism infrastructures cannot solely be repurposed for the domestic market, and that if one wishes to address the disruption to local economies, land dispossession and environmental degradation, coordinated transnational collective action is required.

Mowatt's (Chapter 6) detailed account of the banality of white nationalist tourism drew a similar structural critique of the socialising tourism concept and the tourism industry more broadly. Following the trail of Dylann Roof's white nationalist road trip, Mowatt positioned this "coming of age" narrative within tourism and travel literature. By doing so, Mowat offered an ideological critique of tourism literature, suggesting that it too often ignores these more problematic and large structures of society that work within and through tourism. Such violent and troubling tourism practices like Dylann Roof's road trip may not fit neatly into the dominant managerial, community-based and sustainable tourism discourses; such analyses are, perhaps, too unsettling for some. Nevertheless, tourism as part of the broader society enables this "banality of evil", and as a result, Mowatt suggested that socialising tourism must therefore offer safeguards against such tours that support historical and ongoing expressions of white supremacy and structured racial injustice.

As if responding to the critiques of Bianchi and Mowatt, Kajihiro's (Chapter 8) work with the Hawai'i DeTours project is an exemplary attempt to both dismantle capitalist ideologies that shape tourist social relations and decentre naturalised understandings of history, tourism and places. Kajihiro described DeTours as a practice of "unsettling the settler landscape" by taking people to sites of military occupation and settler colonialism. "Unsettling" is a necessary component of critical pedagogy and decolonial political education for Kānaka 'Ōiwi in Hawai'i. Like Bianchi and Mowatt, Kajihiro also identified the limits to such small tours in the face of ongoing militarisation and colonial occupation, but the story does not end there. As Kajihiro wrote: "we engage in small actions to hold a space within settler societies, even if only momentarily". This creative defiance requires fostering new relationships between participants of tourism and the local peoples and places visited. For Kajihiro, socialising tourism may mean being accountable and responsible for these problematic ideologies, structures and histories as well as building new connections with one another. From unsettling to dismantling, Benjamin and Dillette's (Chapter 7) narrative ethnography explored how the structures of colonialism and a cycle of normativity constrained and limited their own radical potential in tourism scholarship. Their personal reflections as two early career academics revealed how they have been socialised to exist and perform within capitalist structures that permeate tourism academia. Tourism scholarship has a difficult time engaging with themes like those identified by Mowatt (Chapter 6), in large part due to the hegemonic ideology that frames tourism academia itself. And echoing Kajihiro, Benjamin and Dillette concluded by suggesting that the creation of relationships outside of academia is a critical first step in dismantling traditional hegemonic ideologies of tourism academia and opening up new possibilities for alternative tourism futures.

Building alternative tourism futures: multiple scales, multiple logics

Section III explored how such alternative tourism futures may be possible. Many of the contributions throughout this book start with the premise that the endless capitalist pursuit of profit and growth at the expense of local community well-being, human and non-human lives, and the environment has resulted in a socially and ecologically unjust tourism industry. What we have learned about the concept of socialising tourism is that the creation of alternative and possible futures must happen, and is happening, at the intersection of multiple scales and contexts. Section III began with an analysis of post-disaster Fukushima, where Yamashita (Chapter 9) and Doering and Kato (Chapter 10) invited readers to reflect on how local communities reengage with tourism and rebuild possible futures in an environment devastated by a triple disaster: earthquake, tsunami and nuclear fallout. Focusing on the new modes of tourism that emerged following the Great East Japan Earthquake, Yamashita (Chapter 9) introduced the term "public tourism" to conceptualise a form of tourism where the public and tourism non-governmental organisations (NGOs) work collaboratively with communities on public issues, especially in areas devastated by natural and man-made disasters. Yamashita demonstrated that by developing capacity to engage with others, supporting community well-being and establishing better civic society-public connections, tourism can be repurposed to serve the public good, centring the citizenry and people rather than profit.

Several contributions mentioned that building better futures requires greater attention to human relations with the non-human world (Bigby & Jim, Chapter 2; Kajihiro, Chapter 8). Contributions by Doering and Kato (Chapter 10) and Kline (Chapter 11) further developed the idea that greater attention needs to be paid to the non-human worlds that comprise the "public good" and "social justice". Situating the human-environment relationships at the centre of their analysis, Doering and Kato used an ecohumanities perspective to show how communities creatively reconnected with contaminated lands and seas in post-disaster Fukushima. They argued that if we ever hope to build more ecologically just tourism futures, healing, partial recuperation, rebuilding and rethinking tourism requires a more concerted focus on human and non-human relations. They showed how even within capitalist-driven disaster and recovery, it is nevertheless imperative to find ways to maintain and illuminate peoples' connections with their lands and seas. If these human-non-human connections are not maintained, the benefits of entering into a post-capitalist world will remain limited. A similar sentiment was shared in Kline's (Chapter 11) contribution where she introduced the ethical possibilities offered by a post-humanist approach to tourism. Kline argued that the capitalist framing of tourism has led to an ideology of human exceptionalism that has enslaved and oppressed our animal kin. Kline suggested that applying socialist principles of community, egalitarianism and justice within a post-humanist tourism philosophy could be more conducive to the kind of thought and action needed to disrupt tourism's capitalist power structures. This is essential to dismantling the many forms of exceptionalisms and hegemonic hierarchies that lie at the core of tourism's social and ecological injustices.

The final two chapters directly addressed the core tension of sustainable tourism today: growth. Chassagne and Everingham (Chapter 12) introduced the concept *Buen Vivir* to shift the priorities of tourism policy and practice away from growth models focused on tourists towards an acknowledgement of sociological and environmental limits and well-being of host communities. There are two important points for the concept of socialising tourism that emerged from this conversation. First is the broadening of what is meant by "host" or "local" community. Echoing earlier chapters that explored Indigenous traditional knowledges (Peters & Lambert, Chapter 1; Bigby & Jim, Chapter 2; Kajihiro, Chapter 8), Chassagne and Everingham argued that from the perspective of *Buen Vivir*, the "local community" is already inclusive of the non-human world; the social and ecological are inseparable. Second, they showed how prioritising community needs also means respecting planetary ecological limits. This broadened definition of the local community to be inseparable of the non-human world and inclusive of both local and global scales provided a welcomed expansion and clarification of what is meant by the term "community-centred" in the socialising tourism framework.

The final contribution by Fletcher et al. (Chapter 13) also sought to rethink the community-centred focus of the socialising tourism concept, arguing that the erosion of the current capitalist system will require translocal initiatives that may transcend community-level actions. Aligned with Bianchi's (Chapter 6) call for a more radical turn in tourism, the main premise of Fletcher et al.'s contribution is that to achieve genuine sustainability will require a transformation of the larger structures and systems that comprise the global tourism industry. This leads them to ask if a focus on "community-level" initiatives is enough to enact this kind of change. The examples offered in their chapter demonstrated that ultimately it will take a diverse range of action in different contexts and at different scales to socialise tourism and undermine the dominant capitalist framings of the industry that encourage unlimited growth and expansion. We agree and would further argue that in addition to multiple scales, the chapters presented in this edited volume also indicated that multiple logics and world views – ecohumanities, Indigenous traditional knowledges and values, critical pedagogy, post-humanism, the differently defined meanings of "community" in *Buen Vivir* – will also play an important role in unsettling and dismantling many of tourism's hegemonic structures and world views.

Where to now?

Socialising tourism as an avenue for critical thought

The diverse range of contexts, concepts, scales and differing points of tension and attention presented throughout the chapters is indicative of the broad range of methodologies, world views and actions required to foster radical transformation. No one chapter or idea alone could affect the kinds of changes required to confront and counteract the social and ecological devastation the world is facing today. What is evident from this book is that socialising tourism is not necessarily a call for a new paradigm; there are already more than enough paradigms and "turns" in tourism scholarship. But what we do need are new words, phrases and vocabulary or old terms used in new ways that can provoke new action and lines of thought. Socialising tourism may just offer such a term for tourism, providing a certain structure of feeling, ethos, thought and action that can help guide us towards alternative tourism futures.

What is socialising tourism? What work can it do? What are we proposing in the end? We have come to approach socialising tourism as an avenue for thought and action. An avenue is a way, a path, an opportunity, a possibility, and this is the sense of socialising tourism that is experienced and explored throughout this book. Socialising tourism may not (yet) have a clear programme to follow, but there is an ethos predicated on values of respect, collaboration, building and maintaining connections through relationships, an attentiveness to silences and opening spaces for marginalised voices, opinions and writings. It is an approach underpinned by community-led and Indigenous values, ways of doing tourism and approaches to rebuilding lives in the face of environmental ruin and radical colonial disruption. The socialising tourism ethos and action can serve as a methodological tool to inspire others to pick up the challenge of doing, thinking and creating spaces for building better tourism futures. Central to this process is remaining acutely attentive to the critical role of speaking truth to power and calling out ongoing processes of colonisation, oppression, violence, discrimination and unequal structures of capitalist political economies that result in unjust tourism practice and scholarship.

Socialising tourism for achieving justice

This book's title directs us to address the issue of social and ecological justice and draw out some threads from the book chapters to analyse our capacity to socialise tourism for the achievement of justice. From First Nations' perspectives at the beginning of the book, we have been introduced to "Welcome Ceremonies" that work to make the stranger socialised in the local community and to their ways and values. To the end of the book, we have been alerted to the daunting nature of the challenges we face in "eroding" capitalistic forms of tourism, acting at all scales and in collaboration. We opened the book with guiding questions concerning the meaning of socialising tourism, why it is important to socialise tourism and preliminary thoughts on how we might go about it. In this conclusion to the book, we turn our thoughts in a more focused way to issues of justice, particularly social and ecological justice.

Speaking of justice, Mowatt (Chapter 6) confronts us with the multiple and complex ways that touring and tourism supports ongoing injustices not in some tangential way, but rather in a deeply embedded way such that white supremacy and hatred can flourish in the plantation gardens where picnics, festivals and weddings are held to this day. The foreword by McLaren draws our awareness to the impending impacts of climate change that almost make the efforts to draw together a book such as this seem futile or even delusionary. The forces of the one (racist supremacy ideology) are related to the forces of the other (climate vandalism). They are connected via a web of dynamics such as imperialism, enslavement, oppression, domination and lack of care beyond the selfish, grasping self. What are we to make of these overwhelming forces, which at times seem pervasive, unstoppable and so dangerous that they represent forms of existential threat at all levels? These issues must be considered across all levels, from a local, micro context to a planetary level or even galactic lens, as we take war and colonising development into space while simultaneously continuing to poison the home that has sustained us for so long.

With this backdrop of significant injustice, these chapters also reported on the use of the facilities and capacities of tourism to overturn important ongoing social and ecological injustices. This includes Chapter 2 reporting on the use of toxic tours to narrate the experiences of colonisation, mining and pollution that has left social and ecological legacies. These tours accomplish multiple benefits, including keeping the custodians of place connected to their lands and waters, empowering them in creating new futures and securing support from the tourists who are changed by these tours. Similarly, the DeTours in Hawai'i act as a practice in developing a decolonial form of tourism by forefronting the illegal occupation of Hawai'i and its usurpation for military and tourism purposes and undermining its hegemony.

As many of the contributions of this book attest, we completed this edited volume in the context of an unfolding global pandemic of unprecedented proportions. In returning to the present moment and the current COVID-19 crisis, we would be amiss to overlook the fact that these dynamics of inequality and injustice continue to play out before our very eyes in rapidly evolving and unique ways that tie together tourism, justice, privilege, vaccines and health. As we write this conclusion, two new terms have entered our lexicon: "vaccine privilege" and "vaccine tourism". Vaccine privilege refers to the ways in which the wealthier nations have so far been able to monopolise the vaccine against COVID-19 for their own populations at the expense of less wealthy nations and communities, despite the World Health Organisation (WHO) calling for vaccine equity and creating the COVAX initiative to help facilitate this. Vaccine tourism is a related term and phenomenon that describes the deliberate travel by people from places where they are unable to access the vaccine to areas where they can obtain the vaccine. The rise of these two phenomena point not simply to the challenges facing governments and health organisations in addressing this crisis, but unequivocally highlights the underlying structural inequalities and injustices embedded in societies across the globe – while also implicating tourism, movement and access to the means to travel as a mechanism for enabling the privilege (see Loss, 2021). As stated by WHO Chief Tedros Adhanom Ghebreyesus: "I need to be blunt: the world is on the brink of a catastrophic moral failure – and the price of this failure will be paid with lives and livelihoods in the world's poorest countries" (UN News, 2021, n.p.). This moral failure is due to a lack of care for others and a sense of entitlement overcoming any sense of responsibility. Vaccine tourism is an expression of global community that lacks due care and a sense of global responsibility that is an endangerment to our collective future. However, as Chapter 3 demonstrated, tourism can also be an implement to nurture care and caring capacity. To activate justice in tourism to secure more just futures, we must move beyond "moralising" and admonishments; socialising tourism must entail a fostering of our sense of interdependency and mutual entanglements for responsibility to face a future where we either cooperate or we perish.

A research and action agenda

Even with all of the different approaches, perspectives, critiques and potentials that have been explored in relation to socialising tourism and presented in this book, a clear concluding message is that there remains so much more to be done. As highlighted earlier and owing to a number of different circumstances, many important voices and communities were not able to be included in this book. Climate change is undoubtedly a theme at the centre of all of our collective experiences and concerns not only in relation to tourism, but in the innumerable ways it is tied to all aspects of our humanity, our relationships to the living world and our interconnected futures. While climate concerns and human-environmental relationships were explored to varying degrees in some chapters, more direct engagement with the topic of climate change, its full repercussions and its relationship to socialising tourism is research that is urgently needed. Additionally, voices and perspectives on tourism from the LGBTQI+ communities are also important to explore in greater depth how socialising tourism can work to support concerns related to gender, sexuality, identity, rights and equity, among other related themes.

While this book strove to highlight a number of philosophies and perspectives from diverse corners of the globe – aiming for representation from both the Global North and Global South – paradigms and world views from other diverse parts of the world are welcomed and are critical in advancing this work. In addition to diversity of communities, cultural world views and host geographies, the concept of "socialising tourism" would also further benefit from critical engagement and analysis with the diverse spectrum of tourism types that exists. As pointed out by Yamashita's work in Chapter 9, this range encompasses everything from dark tourism, to online tourism and to volunteer tourism as well as innumerable local and Indigenous understandings of "tourism", as highlighted by Melubo and Doering in Chapter 4. New subsets and fields of tourism continue to proliferate and develop, as seen with the recent appearance of "vaccine tourism" as a product of both the COVID-19 pandemic and the privileges exercised by certain countries, certain localities and certain individuals.

Beyond the purely theoretical level, there is an imperative for socialising tourism to critically engage with the on-the-ground development and implementation of tourism as well as the teaching of tourism at all levels. There is both great necessity and potential in having this "avenue for thought and action" that is socialising tourism jump off of the pages and manifest in action through tourism activities on the ground as well as in the classroom and through emerging pedagogies and methodologies. Socialising tourism put into action certainly encompasses the re-centring of community values, rights, voices and interests in shaping tourism that promotes the public good. But this trajectory is not the inevitable or predetermined outcome of socialising tourism for all communities, cultural groups or landscapes. As can be seen through the critiques and pushback to the socialising tourism concept in this very book, in certain cases socialising tourism can be seen to be an inadequate, inappropriate or irrelevant consideration in the interplay between communities and tourism. Indeed, in some contexts, particularly considering the massive existential crises unleashed by both the COVID-19 pandemic and climate change, a critical application of the socialising tourism lens might involve the complete rethinking of conventional systems and paradigms. Socialising tourism does not strive to provide a straightforward or one-size-fits-all programme or paradigm for understanding and conceptualising tourism, whether in a theoretical or practical sense. However, by offering an opportunity, a pathway and "an avenue for thought and action" in understanding how communities relate to tourism, socialising tourism helps in opening the door to critical conceptualisations, rethinking, discussions and transformed tourism activities that are particularly needed in this time of uncertainty, flux and transition. With this approach in mind, the future of socialising tourism is firmly grounded in adding more voices, experiences, critiques, tours and perspectives to the mix. But it also clearly requires us to move beyond a myopic, tourism-centric lens to engage with the wider societal injustices that tourism participates in, enables, causes and/or benefits from.

The unfortunate reality, at least at this current time of writing, is that the COVID-19 pandemic is far from being over, both in terms of being an illness that is claiming lives around the globe as well as a phenomenon that is impacting all aspects of our human realities, from movement restrictions to vaccines to school days. One of the many implications of this fact is that COVID-19 still stands to impact and transform our lives – and tourism, in particular – in ways that we are still yet to understand or even imagine (Munar & Doering, 2020). As our global experiences in navigating this ongoing COVID-19 pandemic evolve, so too will our understandings and relationships with tourism. Indeed, it is at this juncture where socialising tourism thinking may come in quite handy as part of a toolbox for traversing and making sense of the present moment, while also keeping an eye to the future. In embracing the values underlying socialising tourism – namely its search for justice grounded in community-oriented ways of seeing the world – the authors of this volume wish that socialising tourism might serve as a helpful anchor in the ongoing journey of understanding, reimagining and transforming our relationships with tourism, and by extension, with one another and the rest of the living world.

References

Bianchi, R. V. (2009). The "critical turn" in tourism studies: A radical critique. *Tourism geographies, 11*(4), 484–504. https://doi.org/10.1080/14616680903262653.

Fletcher, R. (2011). Sustaining tourism, sustaining capitalism? The tourism industry's role in global capitalist expansion. *Tourism Geographies, 13*(3), 443–61. https://doi.org/10.1080/14616688.2011.570372

Higgins-Desbiolles, F. (2006). More than an "industry": The forgotten power of tourism as a social force. *Tourism Management, 27*(6), 1192–208. https://doi.org/10.1016/j.tourman.2005.05.020

Higgins-Desbiolles, F. (2020). Socialising tourism for social and ecological justice after Covid-19. *Tourism Geographies.* https://doi.org/10.1080/14616688.2020.1757748

Loss, L. (2021). COVID-19: Vaccine tourism is developing around the world. *Tourism Review News* (Online). Retrieved 21 February 2021, from https://www.tourism-review.com/vaccine-tourism-setting-off-around-the-world-news11879.

Munar, A. M., & Doering, A. (2020). COVID-19 the intruder: A brief philosophical reflection on strangeness and hospitality. *Hospitality Insights, 4*(2), 5–6. https://doi.org/10.24135/hi.v4i2.86

UN News (2021, 18 January). WHO chief warns against "catastrophic moral failure" in COVID-19 vaccine access. Retrieved 10 February 2021, from https://news.un.org/en/story/2021/01/1082362

Index

Note: *Italic* page numbers refer to figures and page numbers followed by "n" denote endnotes.

For Product Safety Concerns and Information please contact our EU representative GPSR@taylorandfrancis.com
Taylor & Francis Verlag GmbH, Kaufingerstraße 24, 80331 München, Germany

www.ingramcontent.com/pod-product-compliance
Lightning Source LLC
Chambersburg PA
CBHW052121230326
41598CB00080B/3928

* 9 7 8 0 3 6 7 7 5 9 2 5 4 *